How Many People Are There In My Head?
And In Hers?

To my beloved Godfather
with apologies for
the wayward
thoughts.

Jo.

How Many People Are There In My Head? And In Hers?

An exploration of Single Cell Consciousness

Jonathan C.W. Edwards

ia

imprint-academic.com

ISBN 1-84540-072-0
9781845400729

Published in the UK by Imprint Academic
PO Box 200, Exeter EX5 5YX, UK

Published in the USA by Imprint Academic
Philosophy Documentation Center
PO Box 7147, Charlottesville, VA 22906-7147, USA

A CIP catalogue record for this book is available from the
British Library and US Library of Congress

imprint-academic.com

Contents

Acknowledgements

The germ of the idea in this book arose in 1966 for reasons I no longer remember. The growth of the idea into its 2006 form owes much to constructive criticism and useful suggestions from many people. My thanks are due to those listed alphabetically below and many more, to whom I apologise for the omission. There is no implication that any of them agree with my proposal!

Alec Bangham, Horace Barlow, Rita Carter, Paul Edwards, Michael Fisher, Anthony Freeman, Chris Frith, Karen Gilbert, Ian Glynn, Steven Goldberg, Jo Hajnal, Basil Hiley, David Holder, Andrew Huxley, Paul Marshall, Alfredo Pereira, Alexander Petrov, Gregg Rosenberg, Frederick Sachs, Jack Sarfatti, Alwyn Scott, Steven Sevush, Galen Strawson, Richard Templer, Max Velmans, Guiseppe Vitiello, Andrew Whitaker, Semir Zeki.

A specific acknowledgement is due to Steven Sevush, who came upon the central idea in the book independently at almost exactly the same time. We have had an enjoyable and productive dialogue. My thanks are also due to Anthony Freeman and colleagues at the Journal of Consciousness Studies, not only for useful advice, but also for stimulating the development of the idea in the first place.

The Unthinkable?

...These our actors,
As I foretold you, were all spirits and
Are melted into air, into thin air:
And, like the baseless fabric of this vision,
The cloud-capped towers, the gorgeous palaces,
The solemn temples, the great globe itself,
Yea, all which it inherit, shall dissolve,
And, like this insubstantial pageant faded,
Leave not a rack behind. We are such stuff
As dreams are made on, and our little life
Is rounded with a sleep.

William Shakespeare, The Tempest, Act 4, Scene I

When I look in the mirror I see a head, inside which a story appears to be going on, apparently the story of me. This book is an attempt to pin down how this can come about. My hope is that I can reasonably expect my conclusions to apply to a story inside your head as well, although I may never be able to be sure of that.

Shakespeare's suggestion that we are no more than the stories our heads create for us has been familiar to philosophers over centuries, even if some recent philosophers might claim it to be new. What may be newer is a sense that, knowing so much about the machinery of life, we ought now to be able to understand how it is that our heads come with stories, and in particular, who, or what, is listening.

Any attempt to understand our inner realities requires some leaving behind of familiar beliefs about the way things work, some venturing into uncharted waters. The ultimate goal is to make the way the inside language of our brains describes the physical world match with a way for physics to describe that language, so that we can return home with one consistent story. In a sense I have to build a story about stories. It may lack the narrative of Harry Potter or The Tempest but narrative is mostly a coat hanger for other things and I guess that people read stories as much as anything because they are interested in the way the writer reveals the same sort of inside story as theirs, or gives clues to the rules of how our inside stories unfold and interact. These things are very much what this book is about; in particular about a rule that I think we may have got wrong.

Modern science seems increasingly to agree with Shakespeare. There is little doubt that we *are* such stuff as dreams are made on, not just because we are narratives, but because we are illusions. That is not to say that we are delusions; we can still exist, in the sense that anything does, but much of what we think we are aware of turns out not to be what it seems. This much is widely agreed, but in order to see how the illusion works I suggest that we need to consider another idea, the theme of the book, which looks as if it may have been unfamiliar, or at least unapproachable, to all but one other person on the planet:

Each inside story has many listeners.

I am fairly sure that I am writing this for millions of separate, aware 'listeners' in your head, each one receiving a copy of your story, and its sense of identity, but completely unaware of the awareness of the others, each one a single nerve cell. This may sound like science fiction, and you can treat it as such if you like, but my path of enquiry over the last five years leads me to be fairly sure that each cell in your brain is aware separately and that that is *the only sort of awareness you have* (Edwards, 2005). The idea that we have a single experiencing soul was all a big mistake, even if it is central to the illusion that made *Homo sapiens* so successful.

Before getting readers too worried, however, I would point out that the question in the book's title is not to be taken too seriously. There is only one person in my head and one in yours, one in my wife Siân's head and one in my daughter Beth's. A person is the story that we get when all the cells in the brain work together in harmony. No one cell could create this story, nor could it make all the decisions we make. What it may be able to do is listen to the story. I am suggesting that any living being that observes this page, sees the black ink on white paper and the patterns of the letters, may not be a whole person, but one of the many cells that makes up that person. This is not an easy idea. It is the most difficult idea I have ever come across. Nevertheless, it seems to explain things that nothing else can.

We have known for a very long time that our bodies are colonies of cells, although it was only about 100 years ago that the histologist Ramón y Cajal showed that brain cells, while nearly touching, are separate, not joined together (Ramón y Cajal, 1952). Every cell is in charge of the processes within it. Messages can pass from cell to cell, just as they can from me to you, through this typescript, and groups of cells can act together, as in the pull of a muscle, or in a rowing team, but each cell is a separate life packet.

The idea that each cellular life packet should have its own awareness is not new. It probably arose soon after cells were seen in early microscopes. Even before that, Leibniz had deduced in the seventeenth century that individual parts of us should have perceptions, because logic seemed to require it. Darwin thought that single protozoan cells like amoeba might be aware. The idea that each brain cell might be aware was, according to the father of psychology, William James, well known in the nineteenth century (James, 1890). A few people have toyed with it since. E. A. Liberman, a Russian information scientist, suggested in the 1980s reasons why brain cells could be aware, some of which are very similar to those I shall put forward, but it is not clear that he envisaged many cells being aware of the same things (Liberman and Minina, 1996). It seems that in the past it has been assumed that only one cell had the 'full story' of a person; what William James called a pontifical (Pope) cell. Only one other person that I can discover, Dr Steven Sevush, seems to have taken to heart the possibility that many cells listen to the full story and that there is no one central listener (Sevush, 2006). A dozen others may have done so but not had the opportunity to make their views known.

This book is about the reasons why we should consider seriously the possibility that our inside story has many listeners and why we should not be afraid to do so. It may lead into some unfamiliar places, and for some, perhaps many, people this book may remain as if closed, but for some at least I hope the view is worth the climb. That view does not seem to disturb the ordinary science of the brain. In fact it seems to make it much easier to make sense of. It does, however, put a different slant on who we think we are. Many would argue that the crowning glory of evolution is the emergence of a creature that can reflect on its own being. Maybe there is a step further to go, the ability to reflect on our own beings.

Scientific, historical and literary stories are in a sense just various ways of trying to illustrate the rules. It really depends on how you like the balance between reliability and inspiration in the search for insight. This book is scientific, in the sense of being my best shot at finding rules to explain the way things seem to be that should fit with past and future experiments. However, I hope that the mention of science will not put people off, because there will be no equations or arithmetic. The ideas are the sort of ideas that anybody may either see quite easily or not see at all, which may in itself be an important part of the story. Equally, I hope that scientific readers will not be put off by the everyday language, but I make no apology for this. Scientific jargon is usually designed as much to obscure as to enlighten. What I hope for most is that someone reading this will see through to layers of our inside storytelling that I cannot grasp. I suspect that he or she is more likely to be a teenager than a professor of philosophy or neuroscience burdened by preconceptions and the jargon that goes with them.

I admit that the idea in this book may be wrong, even if I cannot discover why. I thought at first that it might be easy to see why it is impossible, but nobody has been able to convince me why it should be. Moreover, one thing that seems fairly certain about the study of the mind, with its inside story, is that most people agree that up until now no other theory makes much sense. However our minds work, they are at least as strange as I am suggesting. There are lots of new ideas about but in my experience all of them can be taken to bits in five minutes.

There seems little doubt that the problem needs to be looked at from a completely different angle. There is something we are getting wrong.

Yet there are suggestions that the answer may not be far round the corner. This is not because of recent high tech brain scans with pretty colours. True, progress is being made, but most recent reports of research on how the brain organises the sights and sounds that make up our story seem fairly similar in principle to what I was taught by Colin Blakemore as a medical student in 1969. The sense that the answer may be near has more to do with people being prepared to admit we have been on the wrong track and to try new lines of attack.

In 1995 the philosopher David Chalmers wrote an essay called Facing Up to the Problem of Consciousness (Chalmers, 1995), which spelled out in simple language just how hard the problems of understanding our inner selves are. A few people already knew this well, but Chalmers made a lot more people interested enough to stop and think. Although Chalmers had his own suggestions for a solution the impact of the article for me was not 'this is the way home folks', but what is sometimes much more useful; 'isn't it time we admit we might not know where we are going and get the compass out'. It was finding that essay on the Internet that whetted my appetite again for a problem against which I had battled as a schoolboy, and admitted defeat.

Another prompt came from Roger Penrose who, in his book *Shadows of the Mind* (Penrose, 1994), pointed out that it is not good enough to say that our minds are just natural computers and that thinking is just something that happens when electric signals buzz back and forward. He suggested some new approaches relating to modern physics. Penrose knew he had not got the final answer, and like many I think he shot wide of the mark, but he and Chalmers were part of a mood of trying new ideas. If our attempts to understand our minds have got so badly stuck it is reasonable to try any approach, however lunatic it may sound, because, if nothing more, it might show where we are going wrong. It was reading a comment on page 372 of *Shadows of the Mind* that made me think I should pursue seriously the idea in this book. Even if I am wrong, I might reasonably hope that if someone can show why there *cannot* be millions of separately aware cells in our heads, we will have moved closer to knowing what and where 'I' really am, if I am in any place at all.

I have debated the issue with scores of people, from neighbours to the most eminent philosophers and scientists. Most find my idea peculiar, and many cannot accept it, although at least some see the logic behind it. The idea is certainly unsettling. The implications for those who like to hold to beliefs may be truly terrifying. At times I have seen something like a mixture of anger and panic in theists' (and therapists') eyes. Religions that could cope with Darwinism with a little adjustment would probably have to pack up and go home if it proves true. For the open mind, I like to think the idea will not seem too awful. The more I get used to it, the more it seems rather amusing, and perhaps reassuring.

However things develop, I like to think the power of scientific debate might reveal the answer even if such debate has been under threat in recent years. One of the things that made me take a detour into study of the brain was the claim by an eminent immunological colleague that scientific debate simply does not exist, almost as if he preferred it that way. In the twenty first century the Internet has provided a more egalitarian medium in which people can communicate without feeling hemmed in by establishment forces and the 'received wisdom'. The debate is happening and it can take us where we want to go, as long as we keep our feet on the ground.

Perhaps now, more than any time in history, with an eerie alliance of pseudo-democracy and Abrahamic religions presiding over unnecessary war and the destruction of the natural world, we need to know what goes on in our heads. I have not written this story to prove an academic point, or even to be deliberately perverse or mystical. It just seems to make sense, and to be worth sharing. I would also like to think that it might in some way help us to know where to look for the answers to questions like who we are, what we should value, what we should hope for, and how to help our children avoid our mistakes.

Definitions

There are a lot of words that can be used to describe what makes me a 'me'; consciousness, awareness, sentience, thinking, subjectivity, experience, feeling, observing and lots more. Each word means something slightly different and can also be defined in several different ways. I want to stick to everyday terms because jargon always spells trouble. When you get to words like cognition[1] and intentionality you can be pretty sure that they have a different meaning in one book from another, and very likely another page of the same book, and little to do with real life.

Two words seem most useful for my purposes. *Thinking* covers pretty well all of what my venture is about. Descartes (Cottingham, 2001) said 'I think, therefore I am'. However, thinking probably includes both being aware and doing something. I shall come to how I think the link between the two may be quite subtle and certainly not one to one: the 'I' that does may not be quite the same as the 'me' that receives sensations. The bit that most people find most interesting and puzzling is the sensory, observing part of thinking; *awareness*. What makes me me in my universe is being the aware observer.

Awareness is sometimes used just to mean the appearance of being alert or responsive. A bee might be said to be aware of a flower simply because it stops to drink there, without implying it has internal experience. I might be said to be aware that a stove is hot if I take my hand away, even if I do so before I actually sense the heat. However, I am going to use awareness to

---

[1] Cognition means knowing, yet cognitive scientists seem to think it means thinking. Intentionality is said to have nothing to do with intent, but some people think it does.

mean having an internal experience without questioning what or who is having it, because that is what most people think the word means.

So the idea is that thinking needs explaining, and that awareness is the feature that makes the thinking mine and is most crying out for an explanation. What is it, where is it, and how does it relate to the brain substance that seems to allow it to be there?

Part I:
The Questions

one

What Needs Explaining?

> To determine ... what Light is, ... and by what means or actions it
> produceth in our minds the Phantasms of Colours is not so easie.
>
> Isaac Newton

The puzzles of awareness may have always been discussed by philosophers but they are not immediately obvious. Despite a schoolboy interest in where my awareness is, it was not until my thirties that I appreciated just how problematic 'what is blue?' is. Not until my fifties did I realise how much we assume about awareness that we would normally regard as physically impossible. Consciousness buffs may feel they have heard all the questions before, but not everyone has even thought of them. Moreover, after reading six million words on this topic I still find something new in each person's version of the basic arguments.

It interests me that in 1995 David Chalmers needed to call us to face up to the hard problems of consciousness. The founders of our physics, Newton and Leibniz, quite clearly understood how hard these problems are. Leibniz devoted a life's writings to them and, as indicated above, Newton was under no illusion that his optics got to the root of our perceptions. Strangely, many biologists and physicists appear to feel that these problems have gone away, or are trivial, yet I doubt Newton's ghost would think so.

Perhaps the key point is that sensations are built inside our heads. Take movement. The sense of movement is not seeing a thing here now, then there later. The flicker of a flame is as much part of now as its brightness. A tune is not one note now, another later, then another. If it were, we would hear no tune. That has to be arranged inside the head, and the problem is to work out how and where.

Awareness implies that one little bit of the universe 'knows' about other bits, but we need a more precise definition. After a year or two of mulling it over I came to the conclusion that the most useful way to define awareness, as a minimum, is to say that it is access to information. Pedants will then say; 'what do you mean by information – or access?'. I do not think it is that difficult. Most people would agree that to say that I am aware of the room around me implies that I have access to information about the room. If I thought that computers might be aware it would be because they have access to information. At root, knowing must involve access to information.

In neurological terms awareness is sensory: it is having an input. Nobody would doubt that but the odd thing is that most theories of consciousness, even those put forward by neuro-scientists, treat it as something quite different; passing messages around. This curious confusion between receiving messages and passing messages around I will have to return to in later chapters.

But what is the difference between having an input and knowing? Although we may say that we do not think balls know they are being hit by bats we need to make sure we have a reason for thinking that. Being aware must, if it has a physical explanation, be about receiving physical influences, even if they are from memories stored in our head. We are aware of the chime of a bell because we are influenced by the sound waves. We are bits of matter, just like cricket balls, so awareness must be a property of matter which goes along with access to information or physical influence. If we want to think our awareness has more to it than that of a cricket ball we need to pin down the features of our awareness that we do not think cricket balls could have access to. This is not necessarily obvious.

This approach I have found helpful in thinking things through and I will enlarge on it as I go along. The key question is whether receiving influences is all there is to awareness or, as we certainly tend to think, there is more to it. There seem to be two major clues. One is about the 'feel' of awareness, of sensations. The other is about the way information is organised.

Sensations are puzzling. To ask 'what are green, large, smelly, melodious and beautiful?' is, as Daniel Dennett implies, the wrong way to put the question (Dennett, 1991), but it still seems to need an answer. Trying to work out what green is probably takes us nowhere for the time being because it does not seem to have a way in. However, I will spend a little time on this problem because it was Chalmers' (1995) account of just how serious it might be that got me chewing at the mind/body bone again. It made me realise how poorly I had been educated in the insights of people like Spinoza, Hume, Leibniz and Schopenhauer, insights that everyone should be taught about at school. Without them the presumption that we understand ourselves better than the Ancient Greeks did because we have brain scanners cooled by liquid helium is very naïve.

Perhaps the more interesting, because it is more useful, feature of our awareness is that it is organised. A rock might 'feel' heat but does not see a bird because it has no lens to focus light. A cat sees a bird but is not aware it is a vertebrate, because it has no language to focus the idea. Being aware is about having information organised in a variety of ways. This fact is helpful because if we want to track down where awareness is going on we need to find places where there is access to many bits of information in a particular pattern. This I see as my lever to open the box with the answers in, although, as tends to be the case in science, I think I had a peek in the box before I found the lever to check what was inside. This aspect is so important that it will have a chapter of its own later.

Moving from just awareness to thinking as a whole, another feature of awareness already mentioned is that it seems to link in to the rest of thinking in the sense that it influences what we do. If I am asked 'do you sense how lush the green of the grass is?' and I say 'yes', I have a feeling inside that my sense of greenness is influencing my speech. Science as it stands cannot make any sense of this and it is difficult to see how one could start. However, one may be able to make more use of the fact that our behaviour also seems to be guided by our awareness of patterns, like the sudden smile on recognition of a familiar face. It seems that awareness allows lots of bits of information to come together *and influence behaviour together.* It might be argued that it could do it bit by bit, as in a computer, but our brains are too slow for that to be an easy explanation. Moreover, we know that in brains, unlike computers, lots of bits of information do come together.

There is a suggestion that the link between awareness and the rest of thinking is something that seems to allow us to do more than a machine. This is an area of major argument. Some have taken this to mean that thinking is outside physics, a means to nudge physics from the side. This I think has to be an admission of failure; physics is no use if you can nudge it, but I suspect that we must accept that this link involves rather special aspects of physics in unexpected ways. How this might be needs some careful thought.

Sensations and qualia

An element of sensation, like greenness or loudness tends to be referred to as a *quale* (plural; qualia). Like most philosophical terms supposed to be precise, nobody can agree exactly how the word should be used, but because it is short and unfamiliar it is a useful tag for emphasising that the problem of 'what are green and loud' is more serious than one might think.

Qualia are central to the way awareness seems to be more than just access to information. Does all access to information come in the form of qualia, whether it is a person, a jellyfish or a rock that has the access? People are bitterly divided over this but I hope at the end I can convince at least some people that it all makes sense if we look at it in one particular way. I suspect that there is a specific answer or set of answers to the question 'what is going on when red is sensed and how does it differ from green?'. However, for the time being it may be too difficult and there is certainly no reason to think it has to be the same under all circumstances.

A quale is, in the popular sense, a perceived, as opposed to an objective, quality: a quality-in-the-eye-of-the-beholder. Some qualia, like beauty, are obviously perceived. For others, like largeness, it might not be clear that there is a difference between a quality of an object such as size in metres and the perceived quale of being large, but a good example is the enormous expanse of the yellow full moon which appeared over the houses across the road as I first started to type this some time in 2003; a true quale, a subjective, seemed size, a 'too largeness'.

My view on awareness and thinking has changed radically over the years, as much as anything because I had not appreciated just how weird and how important qualia are. This, despite the fact that Bertrand Russell and others pointed out long ago how nineteenth and twentieth century physics and neurology completely fail to explain experience and qualia. Russell's apt comment was that the only thing we are ever aware of is the inside of our own head (Russell, 2002). The awful truth of this is worth some exploration.

Many people use the example of red. I will try the taste of cheddar cheese, because it is easier to see how fickle taste is. And the taste of real cheddar cheese freshly scooped from a waxed cloth-wrapped truckle is something special. (The taste of the rubber you get in supermarkets is unrelated but will do if necessary.) The taste of cheddar is not a property of the cheese itself. There is nothing in or on the cheese that is a taste. Cheddar taste is something that comes after a meeting between the molecules from the cheese and some receptor cells in my mouth and nose. You might say that most of us are agreed that a particular combination of molecules belonging to cheddar cheese represents the same taste for most people, and has a taste of cheddar, but this may well not be true.

The clearest illustration of what I mean comes from studies on mice of different strains which have different receptors determined by their genes. One mouse might consider, let us say, Cheddar and Cheshire cheese to be similar, but Lancashire quite different. Another mouse will find Cheshire and Lancashire almost identical but Cheddar quite distinct. This is because the receptor cells have sensors of different shapes in different mice. In one mouse one receptor cell has a sensor that fits molecules from Cheddar and Cheshire but in the other it fits Cheshire and Lancashire molecules but not Cheddar ones. A unique cheddar taste simply does not exist without a unique cheddar receptor.

And the same applies to colour. Just as tastes and smells are not properties of objects but the results of interactions between our senses and objects, so it must be said to be true of redness.

The cheddar taste certainly does not start to arise until the mouth has received a signal. A few molecules of cheese hit a receptor and this produces an electrical signal that passes on to another cell. We could say that the signal from the receptor to the next cell carries the message 'cheddar', (or at least one feature of cheddar). However, to carry such a message it needs a language. Cheddarness could be coded in a sort of Morse code 'brrrpbrpbrpbp' of electrical impulses, to distinguish it from Lancashireness 'brpbrpbrrrrrrp'. However, one would soon run out of codes for all possible sensations and the evidence we have suggests that the timing of impulses in this context has more to do with intensity than meaning. Cheddarness must lie in where and when the signals arrive at the next cell along, in relation to what else is arriving there and when.

The next cell along might recognise Cheddarness as being an input on a particular cell connection, or synapse. 'Brrps on synapse 47 mean Cheddary.' It may also be important that this brrp is timed to come just before or after another brrp coming in to another synapse. 'Brrps on synapse 47 five microseconds before synapse 129 mean Cheddary'. Either way, there is no Cheddarness in the message; Cheddarness is to do with its arrival.

That might seem to end the matter, but it is fairly clear that the cell that first receives the signal from the taste receptor cell is not the cell that carries the awareness we talk about, which also includes an experience of speech, and other things. The cell connected to the receptor cell never deals with messages about sound or touch. It must pass on the signal of Cheddarness to another cell, and probably on through several other cells before the signal gets mixed in with speech. Do all these cells use the same code, so that a brrp on synapse 47 always means Cheddar? I will return to this issue later, and I have oversimplified it here, but it begs a lot of questions, such as how a cell knows which is synapse 47 and how it attaches the meaning of Cheddar to it. And if it is not cells that know it, what is it, because cells are (more or less) all the brain is made of.

The conclusion has to be that the messages between all these cells are not 'Cheddar' but simply 'brrrps' of electricity. We have to accept that awareness arises somewhere in the brain where signals like the taste of Cheddar, the colour red and speech arrive all mixed together in some electrical language, having been tidied up in other cells which do things like turning individual noises into words. But we have no idea what this 'arising' might involve.

Artificial intelligence enthusiasts may at this point say that it is foolish to think of a cell have a message of cheddarness. They would claim that Cheddarness arises through a complex combination of messages passing through certain circuits. The idea seems to be that Cheddarness might be represented by rush hour traffic, Cheshireness by Saturday afternoon shopping traffic and Lancashireness by Bank Holiday Monday. As I will discuss later I see this as both an invocation of the supernatural and plain cheating, but more of that anon.

The bottom line is that the bit of the brain experiencing Cheddarness has to be seen as receiving messages in just the same way as it would if it were a blob of soggy grey matter in a dish only getting information from the outside world from a mass of electrodes acting like telephone switchboard plugs. How the heck does this bit of brain concoct ideas such as Cheddarness? What and how is it 'tasting' when it tastes Cheddar and how does it know it is a taste rather than a colour? The problem is that it must do. Bertrand Russell was right. The taste of Cheddar is a feature of something inside my head just as red is, and not a feature of my hand-painted copy of Balthus's *Card Game* on the wall above my head.

And Russell's view was not just that of a quirky rarefying philosopher. As indicated at the beginning, Newton and Leibniz took it as read. Thanks to Newton, Maxwell, Einstein and Feynman we have made a lot of progress with regard to what light is. The hard problem remains to work out what on earth in our heads the 'Phantasms of Colours' could be.

As I said before, exactly what tastes and colours might be is difficult to tackle straight off and needs to be left for later, maybe for another generation. For the time being the point to be established is that Cheddarness and red are not to be taken for granted. On their own they seem impossible to understand, but looking at them as part of a pattern of sensations may yield some useful clues.

Organisation into patterns

The feature of our awareness that seems to me to be most useful in trying to pin it down is its richness, the fact that it comes in patterns. The existence of such patterns is a major puzzle. How things like red and Cheshireness are sensed together and in a way that they 'belong' to each other has been called the binding problem (for some this is the combination problem, and for some the binding problem is a sub-problem of this problem but I want to keep things simple). How do we sense qualia 'together'?

Traditional brain science proposes that the workings of the brain occur through the sending of electrical signals very much like those going down a telephone line or in a computer. Each of these signals is in one place at one time. Yet we seem to be aware of a whole lot of signals all at once and in a sensible pattern. Moreover, some of these signals are about taste, like cheddarness, some visual, like the red in the picture, some tactile, like the weight of the laptop on my lap, and some auditory, like the faint sound of Siân's pen moving over a sheet of paper. How do even two of these bits of information exist 'together' if they are signals in different parts of the brain?

The extent of the problem is made worse if we think of the nature of electrical impulses in the brain. An electrical impulse in a nerve is not just a flow of electrons, as in a wire, it is a domino effect of lots of tiny local currents setting each other off all along the nerve fibre; a sort of electrical chain gang. It is worse still if we realise that messages in the brain do not generally consist of single electrical impulses but of machine gun-like bursts of impulses. The message is at least to some extent in the speed or timing of the repeated firing.

Under the microscope it is quite hard to see exactly where the piece of information is that might be Cheddarness. And if the brain experiences taste, colour and speech amongst these patterns of flowing charged atoms, where are they bound together? It makes no sense. Using the ordinary physics of electricity binding does not seem to be on the menu. The impossibility of explaining how different bits of information can be 'in the same place at the same time' is so serious that it made William James (1890) give up any attempt to find a physical explanation for awareness and plump for a supernatural 'soul'. (A few chapters later he gives up the soul, which leaves one thinking that perhaps he has pulled a fast one.) The problem has led to a variety of attempts to dodge the issue. The usual argument is that in complex information-processing structures information just does bind together. The argument seems to be that since we know it does bind in our heads that we have to accept that is the way things are. The sensation and its meaning belong to the whole system, not just to the place where each signal is passing.

The problem is not just that this is a cop out, but it does not fit with what we know. If the sensation and its meaning belonged to the whole 'system' you would not need all the connections to send it around the brain, it would already be 'around the brain'. What study of brain damage has shown is that each bit of information only belongs to the place where the signal is. If you cut the output connections it is lost to awareness. That is the way real things work.

Ultimately, I shall return to the idea that 'awareness is just the way things are' and I do not think we can escape, or should try to escape such a statement. Nevertheless, I think we need to look for where there might be an explanation which, rather than being at odds with all that we are familiar with in either common sense or modern science, is actually part and parcel of the laws of physics and neurobiology which we already work with. We need to discover the code for Cheddarness, the red in the picture, the weight of the laptop, the whisper of the pen and even the comfort of company, all at the same time, but we need to look for it in the right place.

An odd thing is that some people, on reading an account of the binding problem like the one above, will, like me, William James, Leibniz (Woolhouse and Franks, 1998) and a number of modern philosophers of mind, see its importance as self-evident. Other people, some of whom have, equally, worked on the nature of the mind for years, see it as a non-problem. I am not sure that I have ever known someone to move from one group to another, as if brains can work one of two different ways and cannot change. More of this later.

Thinking and actions

Most people would say that thinking means making decisions about things we are aware of in a way that may influence what we do next. Some people might include unconscious brain processes under the heading of thinking. I agree that most of the brain processes involved in thinking are outside our awareness: the processes that link the written word soup to a hiss, a hoot and a pop are invisible to me. However, in common with a number of philosophers, I find it easiest to take the word thinking to mean only those brain processes which at least at some point end up making us aware of something. The processes in our hypothalamus that control our body temperature while we are asleep I would not include in thinking.

There are lots of possible complications here. Can one talk of things that we are 'potentially aware of' but did not notice at the time? Can we separate what we think we are aware of from what we remember of our awareness a few seconds ago? These questions highlight the fact that our awareness is not just a picture of all our thinking processes, it is much, much less than that. However, I still think it is useful to define thinking as all the processes which, taken together, at some stage give rise to the story we are aware of.

If awareness is an essential part of thinking and thinking makes us do things then this might help us pin down what awareness consists of, and where it is going on. It would be easy if we could rely on the popular belief that information comes in from the outside world as sensations, we are aware of these, juggle them, and ideas about them, around and then decide what to do. However, there is a lot of evidence that we do very many things before we are aware of thinking about them. We react to painful injury even before we feel the pain.

We start a sentence knowing that our brain has got a sentence ready to say without knowing what the words are going to be. The work of Benjamin Libet (2002) is most often quoted as laboratory evidence that our brains 'decide' to do things before we are aware we have decided, but I am more convinced by everyday experience. I find myself in that situation all the time.

This leads to the idea that awareness may not be necessary for our actions at all. Yet this seems hard to believe. At least one thinks it determines our 'considered' actions. I raise this issue here only briefly, to introduce it as something to think about later because my conclusion will be that there is an important link between awareness and actions but a subtle and complex one. It has given me more trouble than any other aspect of the problem but I hope I have managed to give it a happy resolution. I think awareness must be in the thick of the decision-making processes in the brain but I also think that animals with behaviour like ours could exist without the sort of awareness we are familiar with. But we need to get further in to the story before I can hope to give an idea why.

Are we just mechanical computers?

There are a number of odd things about thinking which seem to set us apart from computers. Roger Penrose argues strongly in his book *Shadows of the Mind*(Penrose, 1994) that this is so. Penrose gives a mathematical proof that something different is going on. I find the terminology difficult in places and one line of his proof seems self-contradictory, but the common sense drift seems straightforward. You cannot build a computer that knows that the rules it uses for solving problems are both foolproof and complete, because no such set of rules can be devised. We seem to have some supernatural insight into what 'must be right'. We seem to know the rules of the universe without needing to refer to what ought to be an infinite sequence of other rules telling us how to check that these really are the rules. The trouble is that I cannot see how we can be sure this is not an illusion programmed into our brains. Do we 'understand' things, or does our unconscious just do a good job of making us think we do?

Penrose's suggestion that our thoughts could not be mimicked by a computer are, of course, close to our intuition. John Searle has also famously illustrated the idea that computers cannot *understand* with his 'Chinese Room Argument' (Searle, 1997). There seems to be no understanding in the long series of responses to a set of automatic instructions that goes on in a computer, and which could in theory be mimicked exactly by a room full of instructions about what Chinese characters to send out of the room when certain Chinese characters come in, based on the meaning of Chinese, in which a man follows the rules but does not understand Chinese. (A whole book has been devoted to Searle's idea, which contains some very good essays but ultimately shows that philosophers cannot agree on what they understand by the word understand (Preston and Bishop, 2002).)

Even if nobody agrees on what understanding amounts to, it does seem to carry with it some rather odd abilities. We seem to be able to cope with the idea of infinity, even though our brains are finite, and to come to conclusions that ought to require infinitely repeated calculations in a finite time. We seem to make observations that should require counting without any counting apparently occurring.

One of the things which seems to me may be giving us an important clue to the difference between us and computers is the ability to use ideas without needing a specific example. We can apply rules to triangles without defining any particular triangle. This seems to be important for our sense of understanding. When we accept the proof of Pythagoras's theorem we include in the proof the idea that we can see that the sides of the triangle do not have to be any particular length to reach the solution. It seems to have the effect of repeating the test on all possible triangles without actually doing so.

It may be that computers can do this sort of thing, but a number of subjective experiences suggest to me that something odd really is going on. From time to time I clear weeds out of a long flower bed of bearded iris, which gets infested with a plant called creeping cinquefoil which has fans of five small toothed leaves. After an hour or so of weeding my actions become automatic and I find that I am seeing an example of creeping cinquefoil in the foreground of my awareness, with which I match up a plant when I find it. However, if I ask myself how many fans of leaves my cinquefoil image has, I can see it has fans of five, but I cannot say how many fans. It has no number of fans of five even though I see it clearly and my mind carries it with it when I go in for a cup of coffee. It intrudes when I look at the table, waiting for a match, but with no number of branches. Nor are the very real fans of five leaves, of which there is no number, in any particular position, neither at one o'clock, four o'clock or nine o'clock. They are just there.

This image is nothing like the after-image that comes after looking at a bright light. It is what Berkeley implied was impossible 'framing in [the] mind ... a triangle which is [quoting Locke] neither oblique, nor rectangle, equilateral, nor scalenon, but all and none of these at once' (Berman, 2000). It cannot be conjured up at will, as Berkeley rightly pointed out, but in my experience presents itself after repeated exposure to examples of a form being searched for. I realise that this may well not happen to everybody, which may be important to later discussion, but it certainly happens to me.

The things one learns from talking to people with brain disorders, which Oliver Sacks (1986) is so good at illustrating, suggest that we need to be very careful about our assumptions about perception. It is not like what it ought to be like. A man may see where the top of his wife's head is but assume that it is his hat. Qualia do not work together as qualities do, even in normal brains. Patterns can be pure patterns with no number of elements. I suspect that has something to do with how we understand.

Another example that haunts me is Seurat's portrait of his girlfriend making herself beautiful (*Jeune Fille se Poudrant*). Seurat paints the picture in coloured dots, as was his style. The dress is a dazzle of primary spots (with some white mixed in: he had to cheat because the official theory behind his painting had some problems.) The result is that the girl's dress is of *no* colour. Because Seurat borrowed colours to play with light and form, there is no colour left for the dress to have. The bodice is dark. Beth thought it was red. Siân thought it was blue. I suspect it was black satin. The skirt is pale and warmer in colour than the background, but not convincingly yellow or pink. The warmth simply makes it seem nearer than the bluer background. It is no colour, yet nothing seems to be missing.

Annoying weeds and Seurat's girlfriend may not have anything directly to do with Pythagorus. However, my suspicion is that to 'understand' why Pythagorus's theorem applies to all triangles we use a similar mental process in which features of an idea can have no value without the features being absent and the idea losing its meaning. I wish I understood computers better. I am sure that there are tricks that can be used to solve problems with computers so that the answer will apply for any value of a variable, any position of leaves, colour of dress or shape of triangle. However, I suspect the computation either ignores the leaf position completely or assigns an arbitrary value to the leaf position. I cannot conceive that a computer deals with a situation where there is a feature of a system which is present, that by definition has a value but which has no such value. Nor that the computer (or a Chinese Room) would be capable of the further layer of 'understanding' of being able to say to itself 'this is an awfully useful trick for solving problems of infinite sets'. Given the number three we cannot conceive of the computer spontaneously saying 'its all very well putting in 3, what colour 3 is it?, but hang on a minute, it doesn't really matter if it is a brown 3 or a yellow 3, I can handle its 3ness without knowing that and add it to 1 to make a 4 of any colour I like'.

Two things do seem to emerge. Firstly, our ability to solve problems that would seem to require an infinite number of examples seems to be associated with mental activities that if not impossible to program in to a computer, would seem to be very difficult. Secondly, these activities are intimately bound in with qualia, patterns and the peculiar way in which these qualia can be both present and undefined at the same time. We need to find a part of physics where cinquefoil can have no number of fans of five leaves.

As pointed out to me by Karen Gilbert, these problems may have something to do with Chomsky's analysis of language. Words have both a physical presence, as a sound, and a meaning. However, our grammar allows us to play around with these and dissociate and re-associate them, rather as I have done describing Seurat's picture. Perhaps the physical language of awareness is a language of relations between things which, within itself, can relate 'sound' to 'meaning' and if necessary substitute one for the other. The trouble is that at present physics does not include the concept of 'meaning'.

Another aspect of our thinking that seems to stretch the capabilities of a computer-like model of the brain is the ability to solve problems that ought to require counting but without any evidence of such counting. We all know that five oranges are five without actually counting them. There are accounts of individuals who appear able to 'count without counting' very large numbers. (Oliver Sacks (1986) describes twins like this.) Most of us cannot but when we recognise a face in a school photograph we must be doing something very similar. We must 'add up' or at least 'integrate' all the features to recognise a unique combination; one person in ten thousand we knew twenty years ago.

This counting without counting deals not with infinity but large numbers. A computer can recognise large numbers, but it is not that easy to see how it could be achieved as quickly as it is if our brains are like computers of the usual sort, because brains are so much slower. If we could find a physical mechanism for counting without counting it would narrow the credibility gap in our explanation of the brain quite usefully.

There are a variety of reasons for thinking that our brains do things that our familiar digital computers do not do. Although not the first to point it out, Penrose (1994) has certainly made people much more aware of this issue. However, there are probably about half a dozen ways in which brains could be different from computers, in their fundamental physics, their mathematical language, their functional dynamics, their determinism, their patterns of access to information, and their 'self-programming' capabilities. Some of these may prove to be blind alleys in a search for awareness. More or less every possible opinion has been given on their relative importance. What I hope to show is that considering awareness as something belonging to cells individually may make things easier to sort out.

two

A Wild Pigeon Chase?

Not everyone thinks that awareness is such a big issue. Some people dismiss these 'hard' problems as philosophical mumbo-jumbo. Most neuroscientists working in the field seem to regard philosophers (and even physicists) as fools, and vice versa, although the philosophers are more tactful. Words like 'silly', 'misguided' and 'nothing could be further from the truth' are banded about. Even within philosophy one is warned that ideas might 'lead to panpsychism' (which I quite like) as if it were witchcraft. As a scientist one gets used to the idea of rigid received wisdom preventing the development of new ideas. What is disconcerting is to find a subject where there are about five groups of people who all think their received wisdom is right and everyone else is wrong.

Am I, like Roger Penrose, just chasing shadows? Are the puzzles that I set out above just part of a wild goose chase? There are those who would say that as long as we know which part of the brain lights up when we sense red there is no need to ask how red comes about and why it is not green. Daniel Dennett has been dubbed 'the philosopher of mind who claims not to be conscious' because he sees the search for qualia as meaningless. Before rushing headlong into an attempt to solve these puzzles, it may be worth reviewing briefly why so many think others are asking silly questions.

There seem to be two sorts of wet blanket argument. Firstly, there is the idea that awareness is so different from anything else that it will be impossible to understand, or at least mean-ingless to try and describe, because we cannot compare it with anything. Moreover, knowing about why and where aware-ness is going on in our heads is not going to change our understanding of how the brain works because that will be covered by physics and chemistry, which do not include awareness. Basically, awareness is not stuff so you cannot say what it is or does. The second idea is the opposite; that when the brain scientists have worked out the connections in a bit more detail the way that the awareness fits with the physics will just become obvious and until then there is no point worrying about it.

I am fairly sure these arguments are wrong. There are reasons for thinking that the subjective nature of thought does not necessarily make it impossible to understand. Lots of things in science seem like that until someone works out how to deal with them. There are reasons for thinking that awareness is tied in with the physics of our actions, even if in a rather odd way. There are also reasons for thinking that working out all the connections in the brain will not make the hard problems go away.

Awareness as an unknowable mystery

There is a popular point of view amongst scientists who fancy themselves as philosophers of science which says that because awareness can never be observed from outside it will never be possible to fit it in to science, which is about things we can observe and measure 'objectively'. There is a common sense thread to this. We could argue that because it is about observing per se, awareness may be indescribable in the way that it would be hard to answer the child who says 'I know a one and a one make a two and a two and two make a four but what is an and?'. We may be looking for something that cannot be 'found' because there is no relevant thing to compare it with.

But we already know we are looking for awareness not just inside the head but inside the brain. There are quite good arguments for putting the awareness we talk about in certain lobes of the brain and even layers within lobes. These arguments may not be cast iron but if we can narrow things down this far, why not further? Moreover, there are certain things we can say about awareness; it is complex, or rich. Saying more may be difficult, but it is not totally indescribable.

More importantly, the idea that science only deals with the observable, independent of the nature of the experiencing observer, is wrong. You cannot separate observables from observers and the relevance of this to physics has become increasingly clear. For two centuries after Newton science only dealt with things that appeared to be 'anonymously observable'. Einstein showed that this is not enough; the world is different for every observer. In a different way quantum physics also involves a specific observer. In fact it becomes doubtful that anything exists beyond the interactions we call observations. The universe is not made of little billiard balls bumping into each other while we are asleep. The subject, the observer, is already part of physics. I am not suggesting that we need the equations of relativity or quantum theory to see how awareness works. However, any convincing explanation for awareness needs a fundamental physical underpinning, and these are the theories that underpin all physics.

Unfortunately, neither relativity or quantum theory seem to say anything about what an observer is, or what goes on inside it, but without it neither theory will work. Quantum theory requires an observer that is not describable by quantum theory. How bad a theory is that? The formation and acceptance of Newton's ideas was driven by the fact that they fit together; they can be understood. As Feynman put it, nobody can understand modern physics, and many people would accept that this may be because there is something wrong with the way we are looking at the ideas. Physics has a huge hole in it and filling that hole may need more than just equations, it may need some brain biology (physicists may need to learn some anatomy!) and discarding of beliefs about reality (which physicists may not admit having). Physics needs a way of coming to terms with awareness just as much as neuroscience does.

I have already touched on the question of whether awareness affects physical events. There is an argument that it cannot because to do so it would force the universe to break physical laws. Physical laws are supposed to explain all events with no room for other influences. The idea that awareness might tweak or nudge the physics of the brain is a very bad idea unless we can prove it essential. Awareness does not seem to be mentioned in physics books, so it is hard to see how it can act through laws of physics. We seem to be stuck. However, the alternative is that awareness *is* in the physics books, but is just not called that. If so, the effect of awareness on the world is not to tweak or bend the rules; it *is* the rules. This is how Leibniz saw it, and, (as William Seager (1995) has put it) Leibniz was no dimwit when it comes to metaphysics.

For those who cannot see how awareness can be doing things at all, I would agree that I have already more or less defined awareness as the non-doing part of thinking; what a neurologist would call the sensory rather than the motor part. It may be better to ask the question 'Are certain things happening in the brain, which affect the world through what we do, so bound up with awareness, that we would do something different without the awareness?'. If there is only one awareness in a head then the answer should be a simple yes or no. On the other hand if there are many awarenesses, it may be more complicated. There are further layers to this. It is often suggested that the neatest indication that processes that have to have awareness alongside them can have a physical effect, is the fact that people write books like this. Unfortunately, I suspect this is not foolproof. The fact that I am writing this book about my awareness might turn out to prove to me that my awareness affects what I do but it may not prove even the existence of my awareness (or awarenesses) to anyone else, or theirs to me. Each layer of the problem needs to be dissected in its turn; I said I would leave this question until later, and I shall.

In summary, I think the nature of awareness is worth chasing. It may not be possible in my lifetime to prove by experiment the nature and number of my awarenesses, or of the awarenesses of things other than me, whether people or rocks. Nevertheless, it may be possible to conclude by sheer logic that some of the alternatives are self-contradictory, whereas others are not. As Popper (1984) pointed out, this is often the closest we get to proving something in science. The casting aside of self-contradictory ideas has guided science since Galileo. The chief obstacle is often sifting what is self-contradictory from what is merely counterintuitive. We need an explanation that is consistent but must not worry if it seems very strange.

Awareness as a self-solving problem

When I am being cynical, which is quite often, I tend to think that brain scientists may not like to discuss what and where awareness could be too much because you only get money in science to do experiments, so solving a problem by argument without doing an experiment is not popular. Scoffing at metaphysics is a way of saying that 'I belong to the club of good scientists' (who deal with data and do not get their hands dirty with things like ideas). That the point of data is to test ideas seems to get lost.

Such brain scientists run into a difficulty when faced with the fact that they are being paid money to do experiments on the basis that they are actually trying to find out what and where awareness is. To counter this difficulty they come up with the clever idea that understanding awareness is not that tricky after all, but it will only become clear how to understand it after lots of experiments have been done on how signals are sent around the brain.

When Roger Penrose chose as a title *The Emperor's New Mind*, he was implying that brain scientists who think thinking can be explained just by working out the connections in your head have missed the point; the Electrical Emperor has no clothes. Our awareness of the world is more peculiar than we think. Our view of the world is about as much like the real world as a map of the Himalayas is like mountains. We see metal things as shiny, not because there is any shininess stuff out there but because we assign shininess to things that reflect spots of light. A bat may hear metal things as shiny because they reflect spots of sound, but we simply have no idea. Working out all the connections between cells in the brain might make it clear how the brain works as a machine but it may get us nowhere in terms of knowing what or where awareness is unless we take the characteristics of awareness seriously. As I have said, the feature I think may be most helpful at this stage is the organisation of sensations into patterns.

Perhaps the clearest example of how simplistic the approach can be is the suggestion made by many that synchronisation of electrical firing of cells may make us aware of their messages.

The idea seems to be that if a thousand cells all fire together their messages all suddenly become a picture in our mind. This is clearly absurd because the messages are still all completely separate and have to be if they are going to be directed only to the separate nerve cells they are designed to arrive at. Synchronising your text messages with your friends does not make anyone see them all together. Non-scientists might think it odd that scientists should be capable of maintaining and agreeing on ideas that are complete nonsense, but scientists themselves should be familiar enough with this.

David Lodge's point

Some people are hostile to scientific explanation for reasons of upbringing and want to keep the study of the mind in the territory of the arts, or perhaps in the social sciences. They want to keep words, pictures and numbers separate. The suggestion is that a physical explanation for inner human life would somehow devalue it. As someone with degrees in art, science and medicine I see this as a pity. For me these realms have much more in common than is often thought. And there is about the same proportion of hot air to inspiration in both camps, just generated through different languages. Ironically, the most exact discipline I have studied is art history; mistaking a Keating for a Rembrandt is more expensive than publishing bad scientific results.

Both arts and sciences suffer from technical jargon and notation: as Einstein pointed out, a major scientific idea is usually as easy to describe in common language as a play. Both work when they provide access to rules within strange or beautiful things. Both fail when lack of clarity makes you relate things to the wrong sort of rules. As David Lodge points out in 'Thinks' (2001) there may be quite a lot of that going on in mind science just now and the arts people may have reason to think they are ahead of the scientists' game. As I will come to later, the science of the brain may have to accept that the rules of words are as real and legitimate as the rules of numbers, but equally, those rules of words will some day need to marry up with what we call physics.

And I see no reason to fear that a scientific explanation will belittle the human condition. When I believe the idea in this book is roughly right, it increases my respect for humanity. No scientific theory will invalidate the insights of Shakespeare or Goya. The arts have never needed to worry. Science is a rich source of inspiration for art, not a detractor. People often think of impressionism as a reaction to the precise photograph-like paintings of artists like Ingres, but Monet's art is based on photography. The fleeting dabs of light that give the impression of sun over water were first revealed by photography twenty or thirty years before 'Regatta at Argenteuil' was painted. Is Impressionism an example of science belittling painting?

I would see a scientific description of awareness as likely to liberate us, maybe even with a rebirth of arts that seem to have shrivelled in the later twentieth century. I do not mean pseudo-art with blobs to look like molecules. I mean a return to believing that Titian and Verdi had something invaluable to say and that there is no shortage of other things to say as long as people have the courage to address the sensitive core of being human. If the nature of awareness was revealed in the way that photography revealed the appearance of an evanescent glint of light on water, the arts could have a field day.

Other raised eyebrows

There are, of course, a lot of people who do not pause to think about awareness, just as they do not pause to think about why bananas are yellow. At least to begin with Siân and Beth probably thought this project as hair-brained as my looking for a wompoo fruit pigeon in the Queensland rainforest (called wompoo because it sounds like someone talking while brushing their teeth). Brain science and physics are for nerds like dad. I would like to carry these people along as well, since the scientific bits of this story are quite simple, just rather surprising. The wompoo fruit pigeon turned out to be spectacular in its green, primrose and royal purple; everyone agreed. The idea of having a million souls could also be quite spectacular.

Lastly, I am amused by those who, judging by the tilt of their head, clearly think that although the mind is interesting it is hardly likely that a middle-aged rheumatologist might have an impact on a puzzle that has lasted for millennia. These people are just wimps.

Assuming we are not afraid to find a scientific explanation for awareness, we need to find some evidence. We need to try to pin awareness down.

Part II:
Towards Answers

three

Grey Cells

How often have I said to you that when you have eliminated the impossible, whatever remains, however improbable, must be the truth.
Conan Doyle: The Sign of Four (1890)

How much do we need to know about the brain to understand thinking? To set up a new theory, you probably need a wide range of knowledge, and that includes anatomy. Physics and philosophy are no good unless you know what the machine looks like inside. However, if you are prepared to take things on trust I suspect you need almost no knowledge of the complicated anatomy of the brain to follow what I am proposing and so I will not go in to confusing details. If you want a clear account by an expert, read *An Anatomy of Thought* written by my old Cambridge physiology supervisor, Ian Glynn (1999). I will outline those few things about the brain that I think are needed to follow my arguments but leave the sceptical reader to explore elsewhere.

The human brain fills the skull above the level of eyes and ears. It is soft, like set porridge. It is made up of cells, which, where they are packed close together, look greyish, as remarked by Hercule Poirot. Some of the cells are not nerve cells. These are called glial cells and are thought of as 'housekeepers', although some of them may be more involved in thinking than we realise. True nerve cells, or neurones, are the cells that make use of the electrical messages we know are needed for thinking to occur.

Each nerve cell has thousands of input terminals, or synapses, carried on branches sticking out of the cell, called dendrites. The cell has one output cable, the axon, which can be very long, in some cells at the bottom of the brain stretching as far down the spinal cord as your stomach. Bundles of axons take up large parts of the brain, which look whitish because axons are wrapped in fatty insulating sheaths like electric flex. Although there is only one axon, with only one message coming out of the cell at a time, the axon can branch thousands of times to pass the message on to many other cells at the input terminals on their dendrites.

The large number of output connections of brain neurones may not be generally realised and this may be important. My guess is that about 100 million nerve fibres bring signals in to the brain. Within the brain there may be about 10 trillion synapses. Each of these must have an input. That must mean that almost all of these inputs are from other brain cells. If the average number of inputs to a brain cell is, let us suppose, 23,256, then the average number of outputs must be 23,256, near enough. That seems to mean that any piece of incoming information that contributes to a sensation is likely to be passed on to a few thousand places at the first pass through a neurone. If the information continues to be made use of it would in theory only take a few more passes through neurones for it to be possible for it to be sent to every neurone in the brain. If I am savouring the redness in my picture I can be pretty sure that many millions of cells in my brain are getting signals indicating that redness is in my field of attention.

The typical neurone with many input dendrites and one branching output axon is not the only possible arrangement. Neurones seem to make use of variations in structure in any way that is advantageous. Although it is probably true that neurones never have more than one axon, it is not true that the cell can only send messages out down the axon. When an electrical signal passes down the axon it can also pass back up the dendrites where it started off. The dendrites have contacts with other cells, some of which receive signals and some of which send out signals. So dendrites can be outputs as well, but it may be true that the output signals from a cell are always in time with an output down the axon. Individual dendrites probably do not send messages on their own without the whole cell sending a message, although in theory they might. Some cells do not seem to bother with one long axon, presumably because they are designed just to send signals back to the same nearby cells that they get signals from. They may effectively use their dendrites as synchronised outputs. It is important to keep these variations in mind but I do not think they make a big difference to the issues I am trying to deal with. The message is lots of independent inputs and lots of 'copies' of one output.

The outer wall or membrane of a nerve cell is of particular importance to the way the cell works. It forms a continuous cover around the main body of the cell and extends as thin tubes around the dendrites and axon like the skin of an octopus. The membrane keeps certain charged atoms, or ions, inside the cell and certain ions outside. It is the opening of 'gates' on pores in the membrane, allowing these ions to flow through, which causes the electric currents that allow signals to be sent from cell to cell. The opening of gates at one spot in the membrane lets the voltage across the membrane drop. This drop in voltage tends to open the gates in the next bit of membrane. This leads to a domino effect with an electrical signal starting at one point in the cell travelling as a ripple along the membrane. In the dendrites the domino effect will die out unless it is helped by more signals further along. If the domino effect is strong enough when the electrical signal gets to the cell body it can carry on right down the axon without dying out. This electrical wave is called an action potential. As well as passing down the axon it may pass back up the dendrites as mentioned above.

At the end of each branch of the axon there is a bulge that sits very close to a dendrite on another cell. These bulges contain bags (vesicles) of chemicals that burst out of the end of the axon when an electrical signal arrives. The chemicals spread across to the other cell (in about a billionth of a second) and affect the gates in the membrane in the second cell either to make them more likely to open or to prevent them from opening.

This point where the electrical message in one cell links through a chemical signal to an electrical message in the next cell is called a synapse. An important aspect of a synapse is that it carries one signal at a time, much in the way that a single connection in a silicon chip in a computer does. Between neurones signals are all quite separate and only vary in how often they come and in that they can have a positive or negative effect on the voltage in the membrane of the receiving cell.

There is a lot more to say about the way nerve cells work, but as far as I know the passage of information from one cell to another always follows this system of separate electrical signals linking to chemical signals and back to electrical signals at each junction. (Chemicals can also diffuse more widely to synapses on many cells. This is important for states of mind like emotions that last over many seconds or longer. I see this as falling under the same basic type of process.) This is similar enough to the way a computer works and if that is all there is to the brain then it would be reasonable to say that whatever thinking is a computer can probably do it just as well as a brain. As may be clear from what I have said before, it does not seem to help us with the idea of access to patterns of information. For this reason a number of people have suggested that cells may communicate in more mysterious ways, through electromagnetic fields or quantum entanglement. However, all the evidence we have is that cells receive signals only through chemical diffusion at synapses and there needs to be a very good reason for suggesting another mechanism which nobody has seen working.

The conversion of a message from one form to another, as here from electrical to chemical and back again, is known as transduction. One of the ideas that Daniel Dennett (1991) is rude about is that after signals coming into our sense organs have been transduced into electrical signals there is no further, 'second transduction'. What he implies is that the electrical signal is not turned into a picture like on a television screen. However, he is wrong in the sense that there are lots more transductions, from electrical to chemical to electrical and on and on. Dennett would concede this, and would probably suggest that it does not affect his real point. However, it needs to be clear in people's minds that many layers of transduction are going on. As I shall indicate later I see this as very important because there is no continuous electrical flow in the brain, it is all intermittent, and I think that poses further problems for those who would like it to be joined up enough to produce one central awareness.

There is, however, a sense in which Dennett may be wrong in his own terms. There is a bit which tends to be skated over in the story in the brain in the textbooks; how all the one thousand or so chemical signals coming in to a nerve cell affect the way the cell sends out its own signal down its axon. This bit is skated over because it is extremely hard to study. Neuroscientists will admit that there is a gap in knowledge, but it is easy to fall in to the trap of thinking that nothing very important goes in this gap other than electrical signals adding up. Thus the simplest story would be that the more signals come in to the cell the more likely the cell is to send a signal out. However, recent research indicates that things are more complicated and may perhaps involve a further type of transduction. I will say no more about this at this point but will come back to it later when homing in on what awareness might be based on.

Neurones in different parts of the brain are designed differently to do different jobs. There are dozens of different areas with different shaped cells with different patterns of connections. All these areas are known to have particular jobs to do but I will only give a brief outline of those aspects of the brain's arrangement that are directly relevant to where thinking might be going on.

The following description of how the brain is laid out can probably be skipped over if you find anatomy confusing or know it already. However, for those who like to get a feel for which way up things are it may be useful. More details are given in *An Anatomy of Thought* (Glynn, 1999).

The brain can be thought of as vaguely like a cauliflower with the stalk at the bottom being the spinal cord and the whole of the top and sides covered in a mass of crinkled folds, the cortex. Just above where the spinal cord stalk joins the brain other smaller 'stalks' also feed in on each side. These are the cranial nerves that bring in messages from the special sense organs like eyes and ears and also send out messages to control muscles in the face. Different nerves come in at different levels and connect to a series of relay stations, but all except the top pair (for smell) come in to the roughly cylindrical area at the bottom of the brain called the brain stem. The nerves from the nose are odd in that they feed in to the cortex at the front.

Attached to the back of the brain stem, a bit like an extra Brussels sprout on the stalk, is the cerebellum, which seems to act as a relay station for controlling complicated movements like playing the piano, eating spaghetti like an Italian, or just walking without falling over. There are about as many nerve cells in the cerebellum as in the rest of the brain, but everybody assumes it does fairly boring jobs. Whether or not pianists like Alfred Brendl or dancers like Margot Fonteyn would agree these jobs are boring is another matter.

Throughout the brain there are patches called grey matter, where there are a lot of cells, and patches called white matter that are made up mostly of axon bundles. Grey matter is present on the crinkled outer cortex, and in patches in the middle of the brain, that are called nuclei or ganglia. Although the cortex is traditionally thought of as the grey matter where we think, largely because humans have a lot of it, the areas of the brain that are absolutely essential for staying alive, breathing, staying warm and, in particular, being 'conscious' in the sense of being awake, are in the brain stem. The cortex is divided by a gap down the middle into right and left hemispheres. Each hemisphere of cortex forms a continuous crinkled sheet but different areas are specialised for different jobs. The back of the brain deals mostly with sensations, with vision right at the back. The front of the brain has more to do with actions and emotions.

Between the brain stem and the cortex, right in the middle of the brain there are also some large collections of cells known as the thalamus and basal ganglia. These connect to and from the cortex in a wide fan of bundles of axons. They also connect to the brain stem. The thalamus seems to be a particularly important relay station. Most signals coming in from the spinal cord or cranial nerves stop off in the thalamus before going on to the cortex. Wilder Penfield observed that at least one thalamus has to be healthy for consciousness. Because birds do not have a cortex but seem to be aware, and might reasonably even be said to think, there might be an argument for saying that perhaps the thalamus is the seat of awareness, the seat of the soul. The cortex might just be a reference library for checking current experience with memory unconsciously. However, looking at a model of the thalamus brings into focus the big problem the ancients had with finding the soul.

The thalamus comes in two parts, a right and a left, and they are quite obviously separate, apart from a narrow connection. It is hard to escape the impression that if the soul is in the thalamus, there must be two souls. The same actually applies to the cortex, which also comes in right and left halves; if it is seen as a cauliflower the right and left halves are completely separate florets with separate stems. However, because the cortex is so folded around it is not so obvious that it is two separate things unless you look at a section through the brain on an MRI scan.

In fact, when looking at the brain, we find that there are two of almost everything. The right side of the brain looks pretty much the same on an MRI scan or down the microscope as the left side. If we are one person, it seems a bit odd that everything in our brain comes in pairs. Moreover, the two sides are about as separate as one could imagine. There is a large flat bundle of axons, the corpus callosum, running between the two sides across the floor of the deep groove in the top of the brain that separates the right from the left cortex. There are also some smaller bundles crossing over, and a lot of crossing over goes on in the brain stem. However, in the middle of the brain, where you might expect there to be a lot of right to left connections there is vertical slit-like tube filled with fluid that almost completely separates the middle of the right brain from the left. The thalamus and basal ganglia on the right are heavily connected to the cortex and brain stem on the right but not to the thalamus and basal ganglia on the left. The cortex on the right stops quite abruptly in the middle and does not run into the left cortex. As will be discussed in the next chapter it is almost as if the brain is designed to allow the human being to live a left life and a right life with signals between the two being an afterthought. Yes, there are connections, but any idea of the brain as a continuous buzzing, thinking ball is quite hard to match up with the facts.

Another thing against the idea of the brain as working as a whole through some sort of generalised buzzing brainwaves is that the only thing that seems to matter about its shape is that the cells have the right electrical connections. The brain might be thought of as just like what you will find under a telephone cable man-hole cover. Each of a mass of wires coming in is joined up by a little connector to exactly the right wire going out, but then all the wires are just bundled up and stuffed in a waterproof canister. All sorts of diseases lead to changes in the shape of the brain and some people are born with very odd-shaped brains. None of this seems to matter as long as the cells are connected up. If a brain tumour interferes with connections or cells die then the mind stops working properly, but apparently not if the brain just changes shape.

One thing that may get forgotten in the discussion of where awareness is, is that whenever the axons coming out of an area of brain are cut so that there is no connection with other areas, the person reports no awareness of the information in that area of brain. Connections are absolutely necessary for information to be accessible, and as far as we can see only the connections I have described above. There might be awareness of colour and shape in the occipital cortex at the back of the brain but this awareness does not belong to whatever controls our speech. There would have to be another awareness further down the line. Both neuroscientists and philosophers may say I have got the wrong end of the stick but I do not think so, I am just stating the brute facts of the problem.

four

Why Patterns Are So Important

This is in a sense where my story begins and ends. Thinking about patterns made me realise that looking for a place for awareness is not so much trying to turn up a nice idea that looks as if it might work, as facing up to a deep logical problem with what has generally been taken for granted and taking the only way out on offer.

The problem of where in the brain our awareness lies has been around for centuries, and the more we know about the brain, the more impossible it seems to find an answer. At one time it seemed possible that this 'seat of the soul' could be the pineal, a small lump of brain on a stalk tucked away above the brain stem, nestling between the folds of the cerebral cortex. We now know that the pineal is unimportant (and it certainly looks unimportant) and an unconvincing place for the soul. However, the suggestion had a powerful logic to it, probably more logic than the modern idea of awareness being in the cortex. The pineal is one of the few parts of the brain that is single, rather than being one of a pair of right-sided and left-sided parts. On the assumption that there is only one soul in one place the pineal might have to do.

Before people used microscopes it may have seemed difficult to know where the seat of awareness was, but it may not have seemed an impossible question. Put another way, to look for the seat of the soul in the seventeenth century was not necessarily a sign of stupidity. The problem became more difficult when it became known that the brain, like other living tissues, is made up of a very large number of cells, each too small to pick out with the naked eye. The real trouble set in when

Ramón y Cajal (1952), probably the greatest microscopic anatomist of all time, showed, using staining with silver, that brain cells are not part of a continuous net, but separate living units with gaps in between. Moreover, the physiologists, like Sherrington, Adrian, Hodgkin and Huxley, showed that the cells operate individually and pass messages to each other by electrical impulses that trigger chemical messages at the synapses where cells interact. By about 1920 looking for a single site of awareness was probably a very stupid idea; it has just taken rather a long time for people to admit how stupid.

The problem is that the implication of what Cajal told us is too awful; that if we are looking for something within the brain that has access to a pattern of information it has to be a cell, because, as I will discuss shortly, there is nothing bigger than a cell which could have access to a bigger pattern. There is no one cell which would seem to be different from all the others, and we know from brain disease that we can lose cells in almost any part of the brain without necessarily seeming to lose awareness. It is just about possible that there is one cell in the stem of the brain where it joins the spinal cord which is absolutely essential to awareness, but nobody really believes this, and it would pose serious problems with information handling if so. The only solution is that each cell is aware separately and that lots of cells are aware of rather similar things.

Since, it seems, this is rather hard to swallow, for the last hundred years or so people have been looking for explanations for where awareness is that involve lots of cells, and which by definition do not make sense. Many scientists have just shut the question out of their minds. Those who do take an interest seem to believe in some form of what I would call the 'busy bits' theory. This is that awareness is in those parts of the brain that are busy at a particular time, even if being busy tends to mean having an output, when consciousness is access to an input.

Busy bits of the brain can be shown up by picking up electrical waves as an electro-encephalogram (EEG). This method tends to show up most clearly which bits of the outer layer of the brain, the cortex, are busy at any one time. The cortex has been a favourite place for 'thinking' since it only exists in animals fairly closely related to humans although, as I mentioned, it is not there in birds.

More recently, activity in deeper parts of the brain has been shown up using fMRI (functional magnetic resonance imaging) and PET (positron emission tomography). The possibility that awareness or experience could be predominantly based in the deeper 'lower' parts of the brain is something which Jaak Panksepp (2002) has championed, pointing out that the deeper parts of the brain go back much further in evolution and are likely to be where experience got started.

Either way, the idea seems to be that awareness belongs to a network of cells that are busy passing around the information we are thinking about. There are a number of puzzles attached to this. One is that we know that bits of the brain must be busy with things we are doing automatically without thinking about them. Walking through the countryside looking at the view we tend to be unaware that our eyes and feet are busily checking that we are not just about to trip over. Why is some information in our awareness and not other? Another issue is that these bits are only slightly busier than other bits anyway. A standard PET scan shows that the whole brain is busy. To show local busy bits you have to subtract general business from slightly more business.

There is, however, a much more serious problem with the busy bits approach. A network of communicating units does not have access to a pattern of information in any meaningful sense. That is what this book is about. Either you see it or you do not!

Let us suppose that awareness is associated with a particular group of a thousand cells in a busy part of the brain. Let us say each cell has about a thousand inputs. In our busy net of cells some of these inputs will be from other cells in the net so the net as a whole might have an input of rather less than a million bits of information. However, neither the net as a whole, nor any part of the net has access to a million bits of information *as a pattern*. The busy net is like a thousand people each of which receives a thousand Christmas cards and puts them on the mantle-piece. Each person has access to information about the arrangement of a thousand cards but nobody and no thing has access to information about the arrangement of a million cards.

It is difficult to see how a single signal can have much of a meaning in a brain since all the signals are much the same. The meaning of a group of signals must lie in their interrelationships – the way they form a pattern. In the case of the Christmas cards there is no pattern formed by the million cards as a whole that anything has access to. There is only local access to the patterns of a thousand in each house. Similarly nothing has access to the interrelationships between signals arriving at different nerve cells in a brain. There might seem to be but there is not. The interrelationships might become clear later if further messages are passed on to other cells, but that is not the same thing.

Thus, for awareness to 'belong' to a busy net of nerves you have to assume that the elements of information passing through each cell can 'be together' in some way we do not understand. We know how and where elements of information come together in the brain; through connections arriving at individual cells. To suggest that they are together in another way is cheating. If there is another way for them to be together there is no need for connections to send them from cell to cell. The idea is self-contradictory.

The fallacy of nets of nerves having access to patterns of information has been fuelled by interest in pattern recognition by computers. Computers can certainly 'recognise' a very complicated pattern. Moreover, people who work with such computers and neuroscientists both talk about the pattern being 'represented' in the net of semiconductors in the computer or the net of cells in the brain. But the pattern is not actually represented, in the sense of being 're-presented' to something that could 'sense' it in the computer.

All that ever happens in a computer is that each of a long series of ons and offs come together with one of another long series of ons and offs in various places, never more than one from each series at a time, and generate more ons or offs. It is as if each of our thousand people only receives a black or a white Christmas card on December 25th and puts it on the mantle-piece for a day and then throws it away. If you have a clever set of rules you can work out exactly which person received a black card and which a white card by asking the people to send more black and white cards around to each other according to the rules and for one person to send you the cards they receive on a daily basis. But the only person that ends up with access to the original pattern of black and white cards is you, an external observer. The Christmas card receivers will never know. By the same token, nothing in a computer has access to a pattern it is recognising.

The way people seem to have tried to get round this is to say that the access to information of awareness is not access by a physical thing so much as access by an 'information processing system'. In this view access to information is not tied down to a skull or a cell or a bit of silicon chip. It is associated with a set of pathways for information, a set of routes down which signals can be passed. It belongs to 'the system' as a whole, which exists in 'cyberspace'.

A lot of ordinary people would say that this is a bit peculiar. Perhaps this is because they realise that in order to know what it is aware of an 'information processing system' needs to have information about its own extent; where it begins and ends. This is a very serious problem because an information processing system is just what a bystander wants to be a system. A computer can be an information processing system. Two computers linked by a file share system could be a processing system. All the computers in the world on the net could be a system. Does my computer become part of the net system when I dial up the net, or only when I send a message, or receive one? Anyone who ever sent a Christmas card could be part of a system. If I sit on a committee am I an information processing system or am I part of the information processing system of the committee? What happens when a committee member changes? The more you go into these questions the more it becomes clear that there cannot be information processing systems (or Chinese Rooms) that know their own extent.

One might think that this 'boundary problem' might be resolved by saying that a system includes all units sending messages to each other in a particular form; electric currents, sounds or whatever. The fact that messages are continually being converted from electrical to chemical in the brain already makes this a shaky concept. It falls apart when we ask how often something has to send a message to belong to the system; once a second or once a decade? Another serious problem is that the meaning of a pattern of information in such a system is vitally dependent on the *absence* of certain messages. There are an infinite number of messages not coming from an infinite number of places for a system defined this way. You cannot define a system by the traffic of information alone; it is like a chess game with no edges to the chessboard. Something other than an arbitrary onlooker has to define the boundaries of whatever has access to a pattern.

There is not even any consensus about what the system boundaries would be for a person; whether it is part of the brain, or the whole brain or includes the spinal cord or the nerves in the arms and legs or the skin and bone, or even all our worldly possessions.

And even if people want to hold on to the idea of a system it does not work because the idea of the system having access to all the information in it does not actually do anything. It does not appear in any analysis of the way computers work. It cannot have any effect on output, which is entirely dependent on the plodding signal sending from one gate to another inside.

These arguments may make people confused, but they probably only do so because the idea that 'systems', whether of human brain cells or computer chips, can have access to patterns is so generally accepted, despite being impossible. The point is very simple. To explain awareness of patterns of maybe 100 or 1000 elements of information we need an individual physical structure that has access to patterns of 100 or 1000 elements of information. That applies (or might; if you know the catch, still hang in for now) to single brain cells. It does not apply to a 'system' of brain cells, each of which has access to different elements.

I have, over a period of five years, asked myself many thousand times why, if this conclusion is so obvious, so few people have said, 'OK we have to accept this and work within it', rather than 'this must mean that something is going on which physics cannot explain'. As far as I can see it is just that the only way out of the problem is the one I have proposed – that every cell in the brain is aware of the pattern of information coming in to it and that is all there is to awareness – and it seems to be too much for most people to accept.

At this point the exasperated reader may say 'this may be a clever argument but the idea that each cell is aware separately is nonsense, I know that there is only one awareness in my head'. The response to this is that nobody can possibly know that. I will go in to this in much more detail later but at this point I would simply mention the experiments of Roger Sperry and colleagues such as Michael Gazzaniga (1998).

Some years ago Roger Sperry and colleagues performed operations for severe epilepsy which involved cutting the connections between the left and right sides of the brain. The anatomy is a bit more complicated but this is fairly much what was done. It may sound barbaric, but at the time there were no drugs that would help these patients and the operation did seem to work.

After recovering from the operation it appears that most patients went back to behaving pretty normally. To begin with it was not clear that their 'minds' were in any sense divided in two. The patients themselves did not think they had become two people. However, when tested under conditions that prevented the right and left side of the brain from sending clues to each other it became clear that they did have two minds. At least they had two independent half-brains capable of answering questions and performing tasks, even if one side could not use words to more than a very limited extent. The left side of the brain would answer questions with normal speech whereas the right side could answer with pointing, but for some tasks was more skilled than the left.

These results are very much what one would expect from other studies of the brain but they raise questions about how many awarenesses you can have in a head and how they relate to each other. It seems that after the operation the left side of the brain is aware, as judged by behaving like a normal person. It also seems that the right side behaves as if it has a separate awareness. The right side may not be good at language but we do not assume that people who cannot speak after a stroke are not aware. Where did this 'extra' awareness come from? The busy bits enthusiasts might say that now there are two patches of business in the brain so there are two awarenesses. That may be so, but one very important point has been brought out by Gazzaniga (1998).

The left hand speaking part of the brain at least, had no inkling that there might be another unit that by all available criteria makes 'conscious decisions' (which seems to imply awareness) in the same head. In fact the left side of the brain assumed it was in charge of the body's movements as it always has been. When the right side of the brain caused the body to do something for a reason the left side did not know about the left side did not seem surprised. It seems that it did not even enter its (half of the) head that something else might also control the body. It assumed it must have caused the action and invented a reason why it did. It fibbed or, as a neurologist would say, confabulated.

So my point is that we know that in a head where there appear to be at least two separate awarenesses, the awarenesses can be expected to be completely unaware of each other's awareness. Just as I can never know directly if another person is aware, a cell could not have any sense that another cell is aware. Moreover, the deep seated belief that I have one awareness and that each person I meet also has one awareness may not simply be something that we learn from social interaction, it may be an actively self-sustaining lie built in to the system.

If the idea of a single awareness is wrong then why should we have it programmed in to us? Surely that would be a disadvantage? In fact, as I shall discuss later, there are reasons for thinking that it may be a very neat evolutionary ploy. It may be the best way of ensuring they all the cells in our brains co-operate and keep the body that provides them with blood supply healthy.

JOIE and SAMEDI

Turning these ideas over in my head again and again, trying to make sure there is not some terrible flaw, I repeatedly find new, and often simpler ways of putting the arguments. The last section shows how I worked my way in to the ideas by considering how impossible the popular approach is. The reader may or may not have been convinced. Many people with whom I discussed the ideas and several reviewers of what I have written said they did not really see the problem. I began to wonder if I was creating a non-existent difficulty. However, it was at this point that I was pleased to discover, following up a helpful suggestion from Paul Marshall, that William James (1890) had come to exactly the same conclusion as I have just done, a hundred years ago. He became so convinced that it was impossible to conceive of a physical site for awareness that he decided to abandon a physical explanation.

Working through the idea I found myself focusing on two related concepts, which for ease of remembering, and to give the Free French a turn, I have given the acronyms JOIE and SAMEDI. JOIE stands for juncture of integrating elements, SAMEDI for simultaneous access to many elements in defined interrelationships. These may sound complicated and obscure, but they are just attempts to pin down the idea of access to patterns in a general, neutral, but watertight way.

As I understand it the working unit of a computer is called a logic gate. It is a site where two signals arrive and one signal leaves. Ordinary computers deal with signals that can be either on or off. That means that what arrives is effectively on or off together with on or off and what leaves is on or off. The gate can be an AND gate for which you need on and on coming in to get on out, or it can be an OR gate for which you get an on out as long as either one or other incoming signal is an on.

The brain does not use gates of this sort. The equivalent of a computer gate is a neurone. Instead of having two inputs it has thousands. It might still behave very much like an AND or an OR gate, such that either all the inputs had to be on or at least one had to be on. Neurones relaying the sensation of pain or temperature from a finger to the brain would do well to act like

OR gates in a computer. Neurones relaying information about muscle tension might use an AND system. However, neurones in the brain could do a lot of other things. They could produce an output signal if every third input in line was on or if lots of inputs were on near the middle of the cell just after inputs were on further out from the middle, or vice versa but not if all inputs were on. They can be 'if this and this but not that or that and that and not this' gates. They can in theory respond to whatever pattern of inputs they like. A more general term than gate is needed.

A general, neutral term for this sort of interaction of inputs is integration. Since a signal tends to imply something with a meaning, which we may want to keep out of the discussion for the moment, it would be more neutral just to call it an element, being any sort of event or process to be treated as a discrete, all or none unit. A neutral term for somewhere in the universe where things come together is a juncture. So a general label that would include both the computer gate and the more complex neurone would be a juncture of integrating elements; JOIE. I could call this an integrator, but JOIE has the advantage that it does not imply any activity other than a coming together. There need be no little green man that does the integrating (at least for the time being).

The difference between the computer logic gate and the neurone is illustrated by the concept of simultaneous access to many elements in defined interrelationships; SAMEDI. By simultaneous I simply mean that there is at least a short time frame during which all the elements are present; they do not have to start and stop at exactly the same instant. A computer gate has simultaneous access to just two elements in what may be the only relationship that two things can have. Each element is the 'other' element, like a spouse. If a neurone has 40,000 integrating elements, as it may, the relationships are as complex as those of a family tree traced back about seven generations. One element may relate to another as 'fourth cousin three times removed through the maternal/paternal/maternal/maternal line backwards and first born line forwards – or something similarly complex. This relationship can be defined by the position of the input in the tree of dendrites that carry the synapses.

SAMEDI for this neurone can provide great richness because each element is now not just on or off, it is something with 39,999 relationships in space and time to other things. The difficulty that most people have with the idea that a computer could be aware relates to the fact that it seems only to be in a position to be aware of an interminable string of pairs of on and off signals. Something with access to signals that form part of a rich pattern is quite different.

Some people would argue that even 40,000 inputs would not provide anything like a rich enough pattern to explain our awareness, particularly our visual awareness. As someone put it 'how are you going to jam blue and everything else into one cell?'. This is a tough one but I do not think too tough, as I will discuss when coming to the language of awareness later. Firstly, forty thousand inputs may provide the neurone with a pattern of up to a million elements over a period of a second. Secondly, there can be ordered molecular complexity in a neuronal membrane way beyond just the number of synapses which means that each element does not need to be seen as representing an identical 'digit'. That does not increase the number of combinations of elements you can have but it means that each element can act like a word rather than just a 0 or a 1. It is rather like the difference between a 'drawing' and a 'painting' program on a computer. Where a painting program requires you to build a picture in an array of identical pixels a drawing program allows you to create a circle with a single click, which can then be combined with other commands in various ways while still retaining its 'circleness'. Thirdly, We know that the state of the cell membrane is not just determined by the signals coming in at that time. It is also modified by what has been received in both the recent and distant past and whether or not the cell has responded by firing.

If we then consider what patterns a group of 100 cells connected in a network has access to, as a group, the answer is that nothing within the network has access to more complex patterns than a single cell. As indicated before it might seem that the 100 cells have in some sense access to a pattern of 4,000,000 inputs, but nothing actually has access to this pattern because there is no place where it is available. There is no point in inventing a conceptual or 'functional' place where it is all available because it cannot operate physically; it cannot 'function'.

This point may be clarified by focusing on the use of the word simultaneous in the term SAMEDI. A pattern is something that is available all at once, rather than a string of elements, available one after another. In a computer two signals arriving at a gate only have a useful relationship, affecting output, if they are simultaneous in the sense that they are operating during the same time period. The timing of inputs at other gates is, in contrast, irrelevant. Timing is only relevant at each gate separately. The same applies in the brain. The timing of each of 40,000 inputs to one neurone will be critical. This is particularly true because the effect of an input at a synapse is usually a brief positive effect followed by a negative effect. Unless inputs are timed precisely their effects will be chaotic. As in a computer, the timings of inputs at other neurones are irrelevant to our first neurone. The timing of any subsequent outputs that may meet later at another neurone will certainly be important but we are not concerned with these, which relate to a separate episode of integration elsewhere, at a time when the first will have vanished. The crunch is that there is no meaningful SAMEDI, no integration of 4,000,000 elements, there is only separate integration of 40,000 elements at 100 separate places.

The alethiometer

A rather nice metaphorical example of a JOIE with SAMEDI is the alethiometer invented by Philip Pullman in *His Dark Materials* (1995). The alethiometer only takes three inputs at a time but because there are many input options in specific relationships on the dial, each of which has several sub-meanings, the working of the alethiometer might be rather like that of a neurone. The instrument is a JOIE that uses patterns, not a computer gate.

Pullman is one of many to link the paradoxes of the mind to science fiction concepts related to modern physics. I find him convincing because he knows where the real mysteries lie, is strong on insightful allegory (adults lose the ability to use the alethiometer instinctively the way children can) but does not get bogged down in physical details that will not actually work. The magic of the alethiometer is not as fictional as it might seem; things just as magical are present in our heads, and I think they have to be individual neurones.

The impossibility remains?

You may, and probably should, still be puzzled by the concept of SAMEDI and how it works. You *should* be puzzled because in a traditional physical analysis SAMEDI would appear to be physically impossible under any circumstances. Certainly William James thought so. He pointed out that just as there is no SAMEDI for 4,000,000 inputs in 100 neurones so there is no SAMEDI for 40,000 inputs in one neurone because the neurone itself is a network of dendrites. There cannot be any one place, in the cell or anywhere else, where 40,000 inputs meet.

This would seem to be fairly obviously true. However, it may not be. It may be that the cell can be considered as 'one place'. Despite suggestions that this might require esoteric features of quantum theory, (which often sound as if they would give access to the same information in lots of places rather than lots of information in one place) as far as I can see it just needs the properties of waves in ordered materials and some confidence about waves that modern physics provides us with. I will come to the waves in cells shortly.

I hope I have now convinced the reader of one good reason for thinking that awareness might be in a single cell. It is the only sort of place in the brain where information comes together and awareness is about having lots of information together. One reason would not be enough to convince me. There is another reason to come. But before that I would like to deal with some aspects of why awareness in single cells might not be quite as absurd as it may seem to some people.

five

Who Says a Cell Cannot Be Aware?

We are got into fairy land, long ere we have reached the last steps of our theory; and there we have no reason to trust our common methods of argument, or to think that our usual analogies and probabilities have any authority.
  David Hume, An Enquiry Concerning Human Understanding, 1748

The idea that a single cell might be aware in any sense may come as a surprise. It seems to surprise a number of eminent people in the field of mind research. I should repeat that I am not suggesting that a single cell thinks. Thinking must involve shuttling information around between many cells. I am simply suggesting, like Steven Sevush (2006), that a single cell can be aware of the story created by the thinking process. It is a listener to the story, or a viewer of the scene.

The idea that a cell might be aware is not new. I suspect that a hundred years ago it was a respectable and widely considered scientific view. Darwin seems to have assumed that the motion of a protozoan (single cell animal) reflects an awareness similar to, if simpler than, ours. Just how narrow-minded we are in the twenty-first century is shown by William James's comments about the idea of a single cell having awareness. To him it was an idea with a long and distinguished pedigree.

Over a century before, both Spinoza and Liebniz argued that awareness is a property of all matter (Scruton, 2001; Woolhouse and Franks, 1998). This is often referred to as panpsychism, a term that may have caused more confusion than insight. The word panpsychism implies that everything has 'mind', or psyche. It may be that in ancient times people believed that rocks and rivers had an inner narrative of thought like ours, but I doubt that the philosophers of the seventeenth century did. Nor do I think anyone would consider that rocks have a mind in the sense of something that thinks. The word pan-experientialism is better if one wants an -ism at all. The reasonable argument is that since we are made of matter and are aware, it is likely that all matter has some sort of awareness. Almost certainly most enthusiasts for this view would see the awareness of a rock as being a totally banal and trivial state. There is just enough of something there, however, to act as a building block for something much more meaningful in the complex structures of living things. My analogy would be that our awareness is like a 3D equivalent of a DVD played into some high tech virtual reality kit, whereas that of a rock would be like feeding a laundry list into the kit. There might be a few flashes and clicks but no more.

I am not quite sure why so many people object so strongly to pan-experientialism, but as far as I can see there is no single reason. People bring all sorts of mental baggage to the subject.

In fact, a search for a guide to what things can be aware or can think in nature is pretty much like trying to follow the proverbial directions to Donegal. As the Irishman said, 'if you be wanting to know the best way to go Mister, to begin with, I wouldn't be starting from here'. In particular, Western religious and political thought has hammered out the concept of the uniqueness of the human soul for centuries. We start from a position of deeply rooted prejudice. The lack of any reliable basis for this is gradually becoming clearer, but still not clearer to everybody.

We tend to look to science and philosophy for progress in understanding of fundamental problems. Curiously, it seems that in many ways the twentieth century has been one of the least imaginative periods when it comes to such questions, at least until the last decade or so. Things may be loosening up, but in popular culture not much further than David Attenborough smiling benignly on an Orang-Utan that can paddle a canoe and saw wood, implying that she might be a teeny bit like us; or dolphins starring in movies. However enlightened these might seem (and I would be the last to belittle Attenborough's contribution to our appreciation of the natural world) they are steeped in the assumption that human consciousness, if not unique, is very nearly so; that a turtle's brain is pretty much blank. Unfortunately, the debate often never gets off the ground because it is not made clear whether the arguments apply to *thinking* or to *awareness*.

The widespread prejudice in this area is clearly in part cultural, but there are reasons to think that it may be more fundamental to the way our brains are programmed. It may well relate to Gazzaniga's confabulating left half of the brain. The implication of Gazzaniga's observations is that there is built in to the left side of our brain a centre for inventing 'the story of me' and of only one me. It, or something nearby, also invents inside stories for everybody else it meets. And it probably makes up stories for animals' heads if they look and behave sufficiently like people, although it seems to do this in much more detail for Siân and Beth than for me!

There must be a genetic basis for such story telling. It might be that the stories themselves are in the genes. Monkeys' genes contain information which seems to tell them that certain types of snake, eagle or cat are likely to attack them in certain ways. One might even think that the stories about what has a self and what the fate of that self is that are found in the bible are in our genes, there to be recognised by prophets. However, non-Abrahamic religions show that it is not the stories that are genetically determined but rather the tendency for the left side of the brain to interpret the world in terms of some sort of coherent story of 'me' and 'you'. I am not convinced that some chaps who thought they had truths revealed to them two or more millennia ago and wrote them in a big book can be relied on to have got it right.

Although the tendency to see ourselves as one being is probably genetically programmed it is interesting to see that it does not become fully apparent until quite late in childhood. There then comes a time in adulthood when story building often seems to stop and the story becomes immutable. Presumably that is what Philip Pullman (1995) means when he plays with the ideas of the alethiometer only being readable by children, and the daemon that declares the person's state of mind becoming unchangeable as adulthood supervenes. Perhaps flexibility is the key factor that distinguishes our stories from those of our closest animal cousins; the ability to relate things in new ways. Maybe this is what allows us to have language, maybe it is also what makes children so vulnerable until flexibility has been tempered by experience.

We are used to the idea that our senses make up stories that are not borne out by later experience. Illusions occur every day, like the sense that we are moving in a stationary train when the train alongside starts up or the sense that there is someone in the room, when it is only a coat on the door. Science is used to ironing out these illusions. However, if the idea of a single 'me' or of a me of any sort is fundamentally wrong it may be more difficult for the interpreter to cope. It may not cope. One might hope, however, that the same tendency of the interpreter to try to find a coherent story of its self, which has over centuries invented myths, may ultimately spur it on to find the ultimately coherent story – the truth about itself, even if it requires a complete rethink.

I will say no more about the Irish roadblocks formed by religious and cultural myths about what has a 'sentient soul' for the time being and focus on scientific issues. The roadblocks are substantial here too. In the scientific community, the size and complexity of the mammalian brain, and in particular the outer cortex of the human brain, is firmly believed by many to be the substrate which allows 'consciousness' to develop.

Baloney, Seamus: try starting from Limerick.

The flaw in the idea of big complex brains being important not just for awareness but also for sophisticated behaviour and the thinking that regulates it comes from a comparison of a bee and a cow. The problem is that a cow's brain contains ten to a hundred thousand times as many cells as a bee's brain but bees seem cleverer. They not only measure distances and directions, but can describe these to their hive-mates. They set up meetings for communication. Their social network is highly developed. They fashion hexagonal honeycombs. That may not mean they know they need six angles, but they have very refined craft-making quality control. Why should bees not be as intelligent, or as aware, as cows? They may well have just as much information in their DNA, just as many genes, as a cow, a cuckoo, a monkey or a human. The things we link with think-ing - dexterity, ingenuity, learning, language – are not well correlated to brain size or the complexity of neuronal net-works. They are out by a factor of about ten thousand.

One might argue that the skills of a bee do not make it aware, or have thoughts. However, behaviour is the only legitimate guide to where and what awareness and thinking may be. We have no other guide. Perhaps what is more striking is just how dim cows are, with such big brains. Our brains may be biggish for our bodies but complicated social patterns exist equally in birds with tiny brains and baboons with big ones, as the term 'pecking order' illustrates. Network size may not be the key to what we are looking for.

There is a further message here for those who see complex-ity as the secret to the mechanics of consciousness. It has been suggested that consciousness is due to an 'emergent property' arising from the complexity of the human cerebral cortex. An emergent property is a property of something complex, with many elements, which cannot be present in something with very few elements. A good example is a wave on the sea. You only get waves when you have many octillions of water mole-cules. You do not get that sort of wave even in the few sextillions of molecules in a raindrop. The argument is that the consciousness of a human might require a billion nerve cells, and that the few hundred thousand or so neurones of a bee will not do, even if they allow complicated social behaviour. I think this argument needs to be supported by some evidence, even if it may turn out to have some validity. We can all see quite eas-ily why you need an awful lot of water molecules to get the sort of wave you have on the sea. It is to do with there being enough mass to overcome surface tension for instance. It is much more difficult to see why you need an awful lot of nerve cells for awareness. You might need them for sophisticated use of memory and the development of a language with a complex grammar but why for awareness? Why shouldn't the basic requirements be there with only one nerve cell?

There is a twist in the above argument. I have used the com-parison of bees and cows to suggest that awareness is not an emergent property arising from the complexity of nerve cell connections. I have used waves as examples of emergence. Some people would argue that the very idea of emergence is suspect, even for waves. Nevertheless, the suggestions about awareness I want to explore later are very much to do with waves being real things. Complexity, or at least the presence of enough elements to form certain types of order, is essential to these ideas. However, it is not the complexity of the connections in the brain that is involved. Ironically, the human brain may be *too complicated* and specifically, too heterogeneous, for any useful emergent property, like a wave, to set up across it. There may be about 10,000,000,000,000 separate signals changing from electrical to chemical and back in a brain in a second. There are reasons for thinking that this is just far too complex and messy to be a basis for a single awareness. This may be as clear as mud, but please hang in for just a bit longer.

It may be useful to scan down the evolutionary tree to see whether we can identify clear evidence of a cut-off for awareness. Chimpanzees and bonobos are the closest to us, sharing 98% of our DNA code. Together with the other great apes it seems that these animals can take on very human habits if taught at an early age. Their facial expressions, contemplative poses and reactions to both inanimate objects and fellow animals suggest that they are sentient in pretty much the same way as we are. They probably do not reflect on their existence in quite the way that some of us do but in terms of sense of identity they seem to be as much or more like an adult human as an infant human or even one under the age of 11.

It is often claimed that man is the only animal aware of coming death. I am not sure. In a comfortable society man may think less of death than almost any other creature. And anyway we do not even know what death is like or what happens next, or how many beings in our head will die. The fear of death comes along with a deep-seated reaction to danger that we see in all mammals at least and which can come in nightmares without any rational thought attached. The experience of the death of others must be commonplace for wild animals, whether on the Serengeti Plain or Hampstead Heath and it is clearly understood. If a bird is born with an instinctive knowledge of a country to which it will migrate, such that it stops when it gets there, why should it not have an internal picture for death, signs of which it recognises and avoids? It is hard to conceive but I think a big part of the problem is our underestimation of the power of non-verbal thought.

What about other mammals? A child free of prejudice from religious dogma regards her dog, cat or pony as just as aware as herself, with as much of a soul. Our cat Figaro put on a pretty good imitation of thinking in our various opinions. Dogs appear to dream. Birds seem fairly comparable in their behaviour. Just as a chimp will fashion a twig to catch ants so a crow will bend a wire to hook meat out of a jar. A bird's brain is not only very small but it has no cortex of the sort often assumed to carry consciousness in man. The idea that fish might feel pain in the mouth has recently surfaced in the popular press, as if it were something one might be surprised about. It seems reasonable to assume that vertebrates as a whole are pretty aware.

Oysters are famously quoted as not seeming to be aware, at least if you are not an oysterman, but they are quite closely related to octopuses, which seem fairly canny. This emphasises how appearances can be deceptive. Although the much more highly developed eye of the octopus is likely to correlate with something close to our vivid sense of visual awareness, the oyster may perceive a wonderful panorama of tastes in the water passing through its gape. In fact it may see something fairly impressive, for, as I have recently discovered, some shellfish have masses of little 'eyes' along the gape of the shell. And try prizing open an oyster and you will feel it fight like mad. We judge the awareness of animals by how they respond, just as we judge and misjudge levels of human awareness, in children, the deaf, or those suffering from Parkinson's disease or stroke. We know that we can be very wrong.

Perhaps the most telling observation is that anaesthetics that take away our awareness have an effect on *Paramecium*, a single cell. Paramecium also responds in a lively way to the environment around it and may be able to learn the best routes to food. Quite apart from suggesting that any cell can be aware at least in a simple way the fact that *Paramecium* can be 'anaesthetised' suggests that awareness may not require connections between cells. The complex neural nets proposed as a substrate for awareness may have little to do with it. The devotees of complexity and emergent phenomena may have to think again. That is not to say that awareness may not require electrical changes in a membrane. *Paramecium* has a membrane with pores for ions that open and shut and create voltage changes just as nerve cells do.

My impression is that the material being fed in to awareness may become more trivial as we go down the scale of complexity in animals but that the stuff of awareness can probably manage quite happily in one cell. Does that mean that awareness is in all cells? What would be the content of awareness for a skin cell or a plant cell? Perhaps not much, but I suspect that the complex and apparently purposeful behaviour of a skin cell reacting to a wound might well be altered by anaesthetics.

Certainly we should have a healthy respect for the potential awareness of one–celled animals. *Paramecium* may not be very clever, but it has to be able to respond to as many things that affect its survival as it can and it is a big cell. A glorious sunset might be no more wonderful than what *Paramecium* experiences. There is an important difference, nevertheless, between our awareness and that of a single cell, which is relevant to the idea of a unique human consciousness. A single cell has an input directly from the environment. Brain cells receive an indirect input, and as I have suggested earlier, this input is a useful but selective map of the outside world. In higher animals that map includes a depiction of the animal itself – self-awareness, which must exist at various levels of sophistication, of which it is reasonable to think ours is the highest.

An interesting difference between nerve cells and single-celled creatures is that nerve cells do not move. In *Paramecium* input from the environment and movement seem to be coupled directly in the cell membrane. The fact that the animal can follow quite complicated paths around things suggests that surface input may be transferred to some sort of 'memory store', perhaps in deeper cytoskeletal structures. However, within the membrane mechanical, chemical and electrical events seem be coupled in such a way that input and output are two sides of the same coin. Nerve cells use the same types of chemical and electrical events for input and output but have decoupled them from movement. Abandonment of movement may allow far more complex patterns of information to be made use of, but it means that the cell must rely on other cells to be fed, kept warm, and protected from predators. I will come back to these ideas in due course.

What about non-living things? If we accept some form of pan-experientialism we have to consider these as well. After all, we now see the brain as just a complicated lump of chemicals. Life force was discarded as an idea long ago. If a warm mushy lump of chemicals like a brain has sentience, why not a hard little lump of chemicals like a pebble? Allowing awareness for inanimate objects might be seen as a problem if we are to maintain that human brains are very unlike computers. However, we need to distinguish awareness that may go with the basic fabric of inanimate things from awareness related to messages which we as onlookers see as meaningful but which within a computer are just lots of tiny isolated changes in electrical activity.

Awareness at various levels need not have anything to do with the meaning of messages passing through something as judged by an outside human observer. Any awareness in a computer would not be expected to relate to the meaning to us of the information being processed, just as the awareness of a carrier pigeon has nothing to do with the message clipped to its leg. We also need to address the issue of domains of awareness. Would a granite rock have one awareness or would there be one for each crystal of feldspar, mica and quartz, or even for each atom? In a computer would each semiconductor element have a separate awareness or would there be some combined awareness. I think there may be sensible ways of approaching these questions when we come to look at mechanisms of awareness. I would just make the point here that to admit that a rock might have some primitive awareness is not to say that computers must be aware of what *we* think they are doing.

So my main conclusion from this chapter is that awareness of some sort or other is likely to be all over the place. This is certainly easier than having to make it emerge from 'system complexity'. Self-awareness and meditation may require sophisticated patterns of input and circuitry, but I do not see that we can be sure that a bee's brain would not do. The sort of language self-awareness might use comes in a later chapter, but I think 'aware-stuff' is probably any sort of stuff; stuff that is in the physics books.

Now I would like to come back to the issue of how patterns might be experienced. I said that nets of cells would be no good as places for awareness because there is nowhere in a net where more of a pattern is available than in each cell in the net. This raises the question of how anything in a cell can have access to the pattern of information coming in to the cell if the cell is a complicated branching structure with inputs laid out along different branches. It would be easy to argue that access to patterns inside a cell is just as impossible as in a net, so we might as well accept that physics cannot explain awareness and open a comforting bottle of merlot.

However, as I started to turn the problem over in my mind in 2001 it soon became clear to me that physics might be able to solve the problem of access to patterns. My thought was not new, several other people had taken the same step, but they all seemed to lose their grip when creating a bigger picture because they wanted a brain's awareness to be single and global. It seemed to me that it was necessary to keep feet firmly on the ground and match up physics with everything I had learnt about the brain when I was at university and when I worked as a trainee in neurology. It took quite a while to see what was important in the physics and what was not. Michael Fisher and Basil Hiley have been enormously helpful in this respect. The good thing is that at the end of the day I have come to the conclusion that the physics that can explain awareness can be explained without any equations or diagrams or strange terminology. Just as Feynman (1985) was able to produce a workable version of the basic ideas of the quantum theory of light without any equations at all (in his popular book *QED*), it ought to be possible to do the same for awareness. I cannot pretend, however, that I can get the underlying equations out of a drawer to order in the way the Feynman could, although I do have some thoughts about what they might be.

six

Waves and SAMEDI

I have made the case for awareness being in an individual cell on the basis of access to patterns of signals. However, I would like to put this argument to one side for the time being and ask how anything at all in the brain can be allowed by fundamental physics to have this access to patterns, or SAMEDI, without limiting the question to individual cells. The reason for doing this is that I think that the physical question gives the same answer for a different reason; it is individual cells and not brains that have the structure that might allow SAMEDI. My reasons have to do with the behaviour of waves.

Although you might not like to think that you, as a listener, a viewer, a sentient being, are a wave, I can feel myself on fairly safe ground in suggesting it, because that is all there is in the universe according to standard modern physics. Waves come in different types and levels of complexity but there is nothing out there that is not a wave or somehow made up of waves. And if you want to be a single thing you really need to be a single wave. If you are not a single thing we seem to be back to square one on the binding problem. If we are going to try and work out whether or not awareness fits with physics we need to accept what physics has on offer. Physics seems to tell us that awareness should belong to a wave.

The mention of waves may ring alarm bells for those who have heard too many times that physicists would have it that things are both waves and particles at once and who find the idea so meaningless that they switch off. They are probably perfectly entitled to do so because, as I mentioned earlier, current physics is more a collection of ideas fastened together with safety pins than a complete explanation of everything. As

I shall come to later, trying to find an explanation for awareness more or less forces us to try and sort this out, and I shall
try later. For present purposes it is probably reasonable to say
that what physics tells us is that if anything exists in the common sense of something being there even if we are not looking
at it, it is a wave.

As I have indicated, the idea of waves being subjects is not
new. As I sifted through the literature it became clear to me
that it has probably occurred to thousands of people independently. Why it has not been more generally accepted I am
not sure but it may have to do with the terrible knots you can
get tied in if you choose the wrong waves.

What is a wave?

I need to make clear what I mean by a wave. There are times
when there may be better words, like mode of oscillation, as in
the swing of a pendulum, or just perturbation, which might be
the safest way to describe light, although on its own it is less
informative. In simple terms waves are patterns of change in
the universe, rather than lumps. They have amplitude; their
strength, size, or intensity. They also often have a cyclical
property called phase. A wave is usually somewhere in a cycle
of up and down, round and round or some other form of continuous repetitive change.

I say usually, because there are types of wave that are not
repetitive. They are solitary pulses, like the shock wave that
follows a bullet or a tidal wave after an earthquake. In fact the
best known wave in the brain is the solitary electrical pulse
which passes along a nerve cell membrane, explained by, and
named after, Hodgkin and Huxley (1952); the 'H-H' wave.
These solitary H-H waves are vital to our inside stories. However, for the time being I want to concentrate on repetitive
waves or oscillations, with cyclical phase. Examples are ripples on water, musical sound and light.

Mechanical waves like pond ripples and sound can be more
complicated than one might think. The wave in a cello string is
a to and fro wave, but the wave in the air in the body of the cello
that makes the loud sound is a push and pull wave. The up and

down wave on a pond is there because there is a push and pull pressure wave in the water underneath. Waves can travel in solids and fluids but the rules are different. Complex structures can carry unrelated waves in different directions; a film of soapy water held in a wire loop can buzz to and fro like a drum skin while the thickness of the film can be affected by different waves which make it change colour in rings or form vortices which make the colours swirl round.

Thinking of light as a wave can be confusing because it does not oscillate in three-dimensional space. Nothing actually goes up and down, back and forth or round and round. It goes through a complete cycle of something every wavelength but it is not a cycle of movement in the ordinary sense. Interestingly, although Newton felt sure light was not a wave that went up and down he realised that it might go through cycles – what he called 'fits'.

Perhaps what confuses people most about the waves of modern physics is what they are made of and whether waves of water can be treated in the same way as waves of electromagnetism (radio waves, light, x-rays). At this point I will simply say that waves should probably not be thought of as being made of anything, even for the sea. A wave is in essence a package of instructions that tells you the likelihood of so-and-so appearing to be so in a particular place at a particular time. It can tell you the likelihood of a quantum of light appearing to strike a screen in a particular place at a particular time or it can tell you the likelihood of the sea appearing to be up in this place and down in another place.

Waves are not in one place

A wave can potentially occupy a space of any size. It is in lots of places at once. Its progress is influenced by variations in any substance it passes through, so it can be considered to have simultaneous access to many elements of information about the substance at any one time. Moreover, the progression of the wave is influenced by the inter-relationships between these elements, as shown by what are called interference patterns. Perhaps the best example is an x-ray beam passing through a crystal of a protein like haemoglobin. The beam is split into a complicated pattern that tells a crystallographer the interrelationships between the atoms in the protein molecules. So waves have SAMEDI for things they go through and the SAMEDI affects how they behave.

But the immediate reaction to this argument is that it is nonsense. For William James in 1890 and for most people now, a wave is not a single thing that could have 'access' to anything. It is a pattern of movement of lots of things. It does not seem to *exist* as one thing and so would not satisfy Descartes. To say that a wave is having access to lots of different things in different places at the same time is a bit like saying that a swarm of bees has access to lots of flowers at the same time. We are back to our problem. Nobody suggests that bees are telepathic so a wave having SAMEDI seems a sham. However, modern physics indicates that it is not. A wave is one thing, not lots of bits joined together. The classic evidence for this is what happens when a single photon of light passes through a pair of slits in a screen – what is known as Young's double slit experiment, as many people will know very well.

If a bright beam of light is shone at a screen with two slits it is easy to assume that some light goes through one slit and some goes through the other. However, when the light coming through the slits falls on a further screen it shows interference bands of brightness and dimness (a simpler version of the pattern made by the x-ray passing through the crystal). These interference bands are exactly like the patterns you get when a sea wave passes through two gaps in a breakwater, and then 'interferes' with itself to produce new patterns of peaks and troughs. All the light seems to be influenced by the pair of slits.

More importantly, if the beam of light is dimmed until at any one time only one unit of light, or photon, is passing through the screen with slits there is still an interference pattern. This proves that light cannot be divided up into some going through one slit and some through the other. It all goes through both, because, in effect, bits of light do not interfere with each other, only with themselves. Light is not a pattern of lots of little things each doing something different. It is a complete pattern, however much you try and divide it up.

Physicists may say that the idea that a wave has SAMEDI is misleading. As I will come to in a minute, I can see their concern and can happily cope with looking at it the physicist's way, but the physicists have some awkward questions to answer in return, as Roger Penrose points out in *Shadows of the Mind* (1994). A physicist might object that you cannot say what a photon of light has access to at a particular time because nothing can be treated as existing in any particular state until it has been observed (i.e. when the photon has hit the far screen). When x-rays that have passed through a crystal and have hit a screen show a pattern we can describe the history of passage through the crystal that would have produced this effect and that history would include all the possible paths the x-ray could have taken. Physicists in general do not, however, seem to allow themselves to pretend they are the x-ray as it passes through the crystal and see what it looks like from inside.

The problem with this is that if we are not allowed to consider what the world would be like from the point of view of some 'wave-thing' tracking through it we are never going to be able to find out what we are, what it is in our head that has a point of view. Physicist seem to assume that the thing with the point of view, the observer, is some big lump of stuff that does not have to fit in to theories about things that are observed. But these theories tell us that everything in the universe is a wave or collection of waves so we seem to be stuck with having to consider what the world would be like to a wave. We have *got to* pretend that we are the x-ray going through the crystal. Physics has laws about what a big chunk of the universe called a human being can possibly know *about a beam of x-rays* but we are interested in what a beam of x-rays, or whatever it is inside the brain that has our awareness, can know *about the rest of the universe* which is a quite different issue.

A popular way to analyse physics problems is to consider things as existing in a four-dimensional space-time and treat time as a dimension that you can track back and forth in, however you like, just like space. Our forward movement in time is seen as a sort of artefact. There are also reasons from psychological experiments to believe that our sense of time is not 'what time is really like'. So it is not unreasonable to say 'looked at from the outside, the idea that a wave has SAMEDI is an artefact of the human concept of time'. Nevertheless, if we want to take the position of the observer itself, the subject, the 'me' any explanation for what is going on would seem to have to be in the time frame that observers have. Thus, it seems reasonable to say that if it means anything at all to say that two things can happen at once then if a wave is a single thing it must be able to have access to things in two places at once.

But I can also allow the physicists to have their point and admit that in doing so I may get nearer to 'reality'. Let us say that it is not good enough to consider a wave as an undivided thing in several places at one point in time, or even a short period of time. Perhaps we have to say that a wave is a whole undivided thing in space-time which should be seen as a 'history' covering several places, or a domain, over a period of time, the lifetime of the wave. We could call the wave's access to information during this history a 'single frame of observation' and perhaps the S in SAMEDI should be 'single-frame'.

This way of looking at it has the interesting implication that to be any use as a self-contained thing our aware wave really needs both to have a limited domain in space and a limited duration. The duration would have to be short enough for 'now' to seem instantaneous – perhaps less than about a twentieth of a second. This may seem a terribly short life for a listener, but I think there will be ways of softening the blow to self-esteem. Moreover, although the idea that our experience comes in lots of very short bursts may seem puzzling, I am not sure it is necessarily in conflict with our experience. The philosopher Galen Strawson (1999) has suggested purely on grounds of introspection that experience comes in episodes of no more than a few seconds. I would simply suggest that it may be easier to find a physical basis if we cut it down to a bit less than that. I would agree that it seems likely that if experience comes in 'frames' of a twentieth of a second then several frames often seem to run in to a 'camera shot' of a few seconds in which our attention remains on the same material.

It certainly seems uncanny. If a wave on the surface of a canal passes in to two side by side tunnels formed by the piers of a wide stone bridge the wave may encounter boats in both tunnels at the same time and break up into diverging ripples accordingly. It would have SAMEDI for boats. At the end of the tunnels where the wave joined up again a lattice of criss-crossing ripples would form which showed that the wave had met boats in certain places in both tunnels at the same time. We do not want to think of something spread out like this as a *single thing* but why not?

It is possible to extend this discussion at length and I am prepared to believe that I have made errors in the way I have put it across. However, it all boils down to the one counterintuitive conclusion that if the strangeness of a single photon going through both of Young's slits means anything it seems to mean that waves have the SAMEDI we need to make our subjective experience appear to be physically possible. Since *nothing else in our understanding of the universe allows this*, this seems something not to let go of lightly.

One might still argue that light is not a thing, since it is not matter. Nor are sea waves real things because they are collections of matter. One may still resist the idea that waves can be realities. However, since we know that matter and energy are interchangeable and that beams of particles of matter will go through both slits in the screen just like light this will not do. Everything there is in the universe seems to be a wave, so we must accept that whatever in my head listens to the story of me really has to be a wave.

The thing that gives me the most confidence that waves are real is the fact that they do come in units (quanta). If they were not real, why should they come in units, which physicists can show are exactly the right size under conditions where the units can be identified? There are all sorts of things we can change in our universe but the one thing we cannot change is the size of a quantum unit. That seems to suggest that it has an existence in itself rather than being something of an onlooker's making.

Although it is impractical to demonstrate it, waves in the sea can be treated as made of quantal units just as much as waves of light. They are just as much single things. This may confuse those who are used to the idea that processes are either 'quantum mechanical' or 'classical'. However, my understanding is that this is not the right way to look at things. As a mode of oscillation a wave that can take up energy will do so in quanta. Sometimes the wave only has a quantum level aspect, as for light. It may also have a classical aspect, as for sea waves. The way in which the quantum mechanical wave relates to the appearance of a classical sea wave is complex and way beyond what I can cover here. But the way individual quanta might relate to classical waves in the brain is probably of little interest because any effect the wave might have on the brain's output will be classical. The quanta are too tiny to matter. Many people have tried to find ways of explaining what the brain does using the behaviour of individual quanta but this seems to me to be misguided. All we need is for there to be a wave that is a real indivisible thing underlying what appears to be going on.

The reader may still feel uneasy but there is another aspect to waves that may help make things fit a little better into a reasonable view of what is going on. Waves divide up into two sorts and some of the differences between these sorts of waves, and the ways in which they inter-relate are likely to be important. These will be the subjects of the next section.

So William James's problem about it being impossible for any one thing in the head to have access to all the information we are aware of at once is no longer a problem in principle. Everything in my head is a sort of wave and waves have SAMEDI for the medium they inhabit. The difficulty now is a practical one. How do we find a wave that can have SAMEDI for the right sort of elements of information about the outside world in a useful code. We need a wave that only has access to this relevant information, about a scene, a tune, an emotion or the three combined. Information about things inside cells like mitochondria and ribosomes where proteins are made, or electricity passing along a nearby axon, or blood flowing down an artery or brain wobbling due to a bumpy road are of no use. The aware wave must be 'tuned' to be oblivious to these.

Fancy ideas about physics have to be consistent with what we already know about how our nervous system works. We can be fairly sure that a lump of chalk at the top of the white cliffs of Dover cannot have an experience of a seascape because it does not have an eye with a lens. It is reasonable to say that nothing in a computer 'playing' chess has anything like our awareness of a game being played. The waves in the semiconductors in the computer's logic gates have information as unrelated to the patterns we find in the world as the lump of chalk. If a wave in my head has SAMEDI for information about patterns in the outside world there has to be a system for getting the information in the right place at the right time for the wave to have access to it. Once I have covered the different sorts of wave on offer it may be easier to work out how this can be.

Two types of wave

Physics has for centuries distinguished two sorts of phenomenon in the universe. The first is matter. The waves of matter, electrons, protons and neutrons, are not like billiard balls, but do behave a bit like spinning or circling billiard balls in two important respects. They come as single units or quanta, and each unit takes up a space, in the sense that two units of the same sort cannot be in the same place at once. So if helium atom has two electrons around its nucleus and another electron comes along, the message is 'sorry full up, no room for any more electrons here'.

The second sort of phenomenon is a traditional wave; something associated with a force, such as light, sound, sea waves and the like. These waves are also based on units or quanta. Photons, which make up the electromagnetic force fields that we observe as radio waves or light are the most familiar force quanta. Sound and sea waves are not normally considered in terms of constituent units because the units are so trivial there is rarely any point in doing so. Nevertheless, it seems that it is equally legitimate to consider sound or elastic waves as built of quanta, technically called phonons.

Force fields differ from matter in one fundamental way. They *do not exclude other waves from their space* but can add up to form stronger waves. This is familiar to all of us. Nobody would ask 'How much light can you fit in this jar?'. You can fit us much light as you like in a jar (until it melts). Because units of a force field do not exclude each other from a space it is much less clear that they should be seen as each having a separate existence. Separate existence is something many physicists prefer not to comment on but when David Bohm and Basil Hiley (1995), in their interpretation of quantum theory, described things as 'be-ables' rather than just 'observables', with the idea of giving them some existence, they made each quantum of matter a beable, but for force fields they made the beables relate to the whole field (and for specific reasons to do with the mathematics). Quanta of force like photons may be not so much separate things as fixed steps of strength by which a field can increase or decrease. (This may help to keep individual quanta out of the story.)

There is an ironic aspect of the way physics divides things into these two phenomena, matter waves and force waves; it is 'dualistic'. Descartes is often dismissed as a dualist who claimed that in addition to the physically real stuff of matter, *res extensa*, there is another type of stuff which does not take up space, which includes light and thought, *res cogitans* (Cottingham, 2001). This is often seen as being some sort of 'ghost stuff' outside physics. However, light is a perfectly respectable non-matter part of physics, so Descartes, at least in this respect, may have been pretty close to the modern interpretation. If awareness is based on a force field then it will have no 'extension' in the sense that it does not exclude other fields from a space. That does not mean that it does not occupy a certain patch of space, but that does not seem to be what Descartes was saying; both light and awareness appear to reside in one place or another, they just do not seem to *take up* a space.

There may be another irony in the relationship between material substance and the two types of wave. Leibniz argued that if things are divided up into their elemental components then many of the qualities we associated with objects cannot apply. An atom of matter cannot be hard because hardness is one part of a thing being difficult to move relative to another part and an atom has no parts that you could move around. Things too small to see cannot be opaque. Modern physics seems to have shown Leibniz to be right. The smallest elements of the universe have no hardness or opacity. Hardness comes with large collections of 'matter waves' particularly where there is cohesion, forming 'objects'.

Paradoxically, cohesion is a force: qualifying as an object may have a lot to do with being associated with the other sort of wave. It is a force field that makes a bunch of atoms hard, gives it shape and determines what sort of space it takes up. This has the strange implication that the things we think of as matter are not so much matter as the force fields associated with arrangements of matter. It is not surprising that there are endless arguments, particularly by philosophers unfamiliar with physics, about the meaning of 'physicalism' or 'material-ism'. It is not that simple.

Force fields come in very different shapes and sizes. Electro-magnetic forces are mediated by radio waves, light or gamma rays, which on a clear day can travel out to the furthest parts of the universe. Force fields can also relate to local, well-defined domains of the universe. In the latter case, as indicated above, the force field and the 'object' it occupies are just two aspects of the same thing. The force fields that we are most familiar with in daily life are elastic force fields. The elastic field that holds a bell together makes a nice example. The chime of a bell is an elastic wave that is a pattern of change in the domain of a par-ticular piece of metal. The wave is a reverberating wave, which occupies the entire bell, and only the bell. The force field should have SAMEDI for every part of the bell and, at least at the level we are interested in, only for the bell.

Waves, perception and apperception

In the context of Leibniz's approach to matter it is worth noting another of his claims; he stated that the indivisible units of sub-stance which he called monads (he had no knowledge of the atoms we are familiar with now, just the concept of indivisible bits of stuff) must have some perception, if indistinct, of the whole of the universe, and a higher level, clear, perception, which he called 'apperception', of a particular domain that is specifically relevant to that monad (Woolhouse and Franks, 1998). By perception Leibniz appears to mean access to information and by apperception he implies awareness of the sort we are familiar with. He also calls his monads 'souls'.

Leibniz's 'perception of the universe' might only imply access to resultants of forces like gravity and electromagnetism in the way that standard classical physics implies. A positively charged particle surrounded by seventy seven negatively charged particles does not in classical physics need to have any information about how many negative particles there are or where they are. It behaves as if all the electrical forces add up to produce one 'resultant' force. However, 'perception' seems to imply access to detailed information and if apperception is the basis of our human perception, as Leibniz would have it, then that would certainly be the case. Leibniz seems to have implied that monads have SAMEDI and that for monads with defined domains the SAMEDI for the domain is in a heightened form. That seems to fit with localised force fields, like in a bell. Leibniz made certain other claims about monads, such as being indestructible, which now seem inappropriate, and specifically to force fields, but I am not suggesting he got everything right!

Waves and understanding

A further claim of Leibniz's that seems insightful is that the indivisible unit or monad progresses through time in harmony with the whole of its perception, and dominantly with its apperception. This is a reasonable old-fashioned way of saying precisely what quantum mechanics says. As a wave moves through space it seems to respond instantaneously to a complex pattern of information about every detail of the section of the universe it could have reached in a given time.

How can it possibly do that? As a number of eminent physicists have pointed out, the most extraordinary thing about the universe is that every bit of it seems to know exactly what to do. When quantum theory was invented somebody forgot to add 'this theory implies that waves really exist and as such have to be seen as unbelievably clever things that can perform infinitely complex calculations on patterns in space time, based on an innate knowledge of the laws of the universe, instantaneously and without any energy requirement'. It is no good to say that a moving thing goes on moving because it knows no better. Even something like light moving in a straight line does so because *it has considered all the possible alternatives* and chosen the 'right' one.

In a sense, a wave *must understand the rules of the universe* since it decides where to go on the basis of what those rules say about its options. If you want a direct line to the rules of the universe, which seems to be the only way round the problem Roger Penrose (1994) raised with human being able to see the answers to problems that computers can never solve, be a bit of the universe, be a wave. Perhaps a wave ought to know when a mathematical theorem is right because it knows the way the universe works. The problem with physics is that it has made the observer outside the analysis of the 'observed universe'. The earth in orbiting the sun is said not to compute its path, unlike a computer-driven satellite. But is it computing in a more fundamental sense, in a way that the human brain has evolved to make use of? Is that why our computing is not just that of a mechanical computer?

The answer to this is probably complex. Like the mechanical computer and unlike the orbiting earth we map things into and out of codes, but the 'sums' done on the codes may still be very different, arising from a universal 'automatic' mathematics. For those who never liked maths this should perhaps be as comforting as it is peculiar. Perhaps schoolroom maths is actually a cumbersome, trivial substitute for an innate ability of the universe, including bits of the universe inside our heads, to do infinitely complicated sums without trying. Schoolroom maths has a practical convenience when you want to write things down precisely but it may seem perverse and contrived to many people because that is what it is. It may not be the way the universe works and not the way our minds work by instinct.

What Sort of Wave Might Be Aware Like Us?

So, if we are looking for some individual thing that might exist, and so satisfy Descartes's claim to exist because he thought, it ought to be a wave. If the option is either a single matter wave, like an electron, or a wave associated with a relatively large chunk of stuff, like a force field, then there are obvious advantages in going for the latter. Trying to get a rich pattern of meaningful messages to a single electron is what made me give up this game thirty-five years ago.

But force fields can have their problems too. Ordinary electromagnetic force fields carried by radio waves or light are unlikely to be much use for thinking because once the waves have been created they either zip off into the distance at 300,000 kilometres a second or, because the field has access to bits of plumbing like arteries, very quickly get randomly scattered or absorbed to produce heat. To have an electromagnetic field in a defined domain you effectively need a transmitter and an aerial and there appear to be no relevant aerials in the brain. What we seem to need is some other sort of wave that stays put but which can extend over an area complicated enough to link up with the electrical goings on in the brain in an organised way and which only has access to the electrical events that we know carry brain signals.

So the suggestion is that we should look for a wave that is a force field in a clearly defined domain somewhere inside the skull, somewhere in the brain. We would like it to occupy a domain that receives a rich pattern of information and we would like the forces involved to have something to do with the way the information is carried. A nuclear force field is not likely to be much good since we would not expect thinking to affect nuclear forces too much! The most familiar local force field is an elastic force field of the type that makes a bell ring. This gives a reasonable analogy to use as a starting point although the complexity of living structures would allow a variety of more complicated force fields to develop.

The analogy with a bell might lead us a bit further. If I tap a good bell, it rings with a single pitch, or a small set of harmonic pitches, all in tune. Because the bell has a well-ordered structure it has a dominant wave mode, which makes it a bell. Because the mode is dominant we can reliably notice it: it speaks. If the bell is cracked, even if it is still in one piece, it may no longer have a dominant mode, it will have many muffled modes. It is still a bell shaped piece of metal but not really a bell.

Perhaps we are looking for a dominant mode of oscillation in the brain that can speak, almost literally, if indirectly. If awareness controls our speech, that would seem to be needed. If so the field would surely need to dominate the domain it fills. However, I am not at all sure that our speaking comes from a single domain, or even the force fields in lots of domains. I need to move one step at a time. For the moment I am simply trying to find a place for awareness. What happens next I need to leave for another chapter. Nevertheless, a domain with a stable dominant wave mode seems more likely to have an awareness in a useful sense than one with a muddle of unstable modes, regardless of the output issue.

And whether or not it is essential to output, the presence of a clear dominant wave for a domain has a useful flip side to it. The wave is clear and dominant partly because the domain has a well ordered structure but also because it has clear boundaries. This means that a dominant mode *only* has meaningful access to information about its domain. The rest of the universe is, more or less, shut out. This would seem to fit with the idea that an awareness is closed in the sense that it cannot share its observing with other awarenesses; you cannot have telepathy. That suits me because I reckon that if telepathy were possible we would have evolved to use it all the time. One thing that my ideas are not designed to explain is ESP.

If awareness needs a wave mode, what inside my brain can have a reasonably stable, dominant mode? Brain tissue is not something that one would expect to ring, or in fact support any very interesting mode of oscillation. There are two problems.

Firstly, the brain as a whole hardly qualifies as a separate physical thing with a distinct domain, or even a distinct function. Within the brain, cells are held together by fibres of collagen and other gluey proteins. Much the same things connect brain cells to spinal cord cells, to peripheral nerve cells, to muscle cells, skin cells and… the whole body. Even the electrical processes going on in the brain also go on in spinal cord, nerves and muscles. It is very difficult to see how the brain can qualify as the sort of separate thing that could have its own wave or be an observer. A subsection of the brain in the form of a 'neural net' will have even less well-defined boundaries.

In response to this argument, that brains do not have clear edges, a number of philosophers have suggested it is better to treat the whole person as the subject, rather than the brain. Wittgenstein tends to be wheeled in here. The idea is that to say that a brain has thoughts in meaningless. Thoughts belong to a person, a human being. To me this is burying your head in philosophical sand. Conjoined twins show us that the idea of a whole human being having thoughts can break down and that it is at least one stage clearer to admit that thoughts belong to brains. Dreams certainly seem to.

Moreover, considering a whole human being will certainly not help us find a wave with the sort of SAMEDI we need. True, a whole human being can have a mode of oscillation. A person involved in bunjy jumping can spin round in space as an individual object, which is a wave mode of a sort. But so can a person with a rucksack containing a take away curry and two portions of fried rice. There are all sorts of levels of being an individual thing because of a way of oscillating and some of them really aren't going to help us understand awareness, which is not manifest through spinning around holding a curry! Two ice skaters holding hands and whirling around each other do not develop telepathy. We need a wave tuned only to the electrical messages of the brain, which fairly obviously needs to be inside the brain itself.

The second problem is that the consistency of the brain is unpromising as a substrate for waves. It is more or less like putty in that it is a mix of solid and liquid components; about the best possible stuff for damping out waves that might set up in it. Again, I would not deny that a brain can support some trivial sort of elastic wave. However feebly, a brain might hum or wobble in a skull. That wobbling would be affected by every part of the brain so the elastic wave would have SAMEDI for the whole brain. But it is not credible that this would be the basis of awareness. The brain has blood vessels pumping through it, sheets of fibrous meninges to keep it place and a lot of other 'plumbing' of no interest to thought. An elastic wave coursing through the brain would be as much use to thought as the cleaning lady in the corridors of the British Library would have to a scholar wanting to read a first edition of Chaucer. I cannot say that a brain cannot support an elastic wave but I think I can say it cannot support a useful elastic wave. Moreover, the heterogeneity that makes the elastic wave unworkable almost certainly makes any more sophisticated wave equally useless.

The brain as a whole, or even a chunk of brain, is no use. It is an aggregate. Whether we think of electromagnetic or elastic waves, the brain is a complete muddle. It is not a muddle if we are considering it as a telecommunications network but as something to support a wave it is hopeless. We need to look for something smaller than the brain, something less of a muddle; the problem with the wonderful complexity of the brain is that we actually need something less complex. Whatever has the dominant mode must have quite a large number of inputs to have access to the richness of our experience but the question is; just how complex do we need the structure to be to have access to enough input?

We tend to think we see several thousand things at once but the psychologists tell us that it may be a few hundred things, or even less, put together in some sort of internal shorthand language. Each of these things may require several elements for their description but it seems unlikely that our awareness needs more than a million or so elements of input per second.

On this basis, the single brain cell may be just about right. In the cortex of the brain nerve cells have something between a thousand and fifty thousand inputs, each of which can receive a signal up to a hundred times a second. The crucial question is whether or not they can support useful wave modes; whether they will fit the bill.

Can we be sure the brain does not have a useful wave?

But before going on to cells let me try this question one more time. It greets me every day and has greeted me every day for at least four years. Surely it would be better to go back to trying to find a way to make a single awareness in a head than to continue with this mad idea about a story with each cell a separate listener? This is not the last time I will come back to it but I would like to clear away one or two more myths before we move on.

A number of people have made attempts to explain consciousness on the basis of fields, chiefly electromagnetic fields. Karl Pribram (1991) has been very interested in the way information in the brain seems to be spread out, rather in the way that it is in a hologram. He has suggested that some sort of global field is set up which allows perception. Others have invoked electromagnetic fields with a less detailed rationale. I shall return to the holographic aspect of thinking suggested by Pribram later, because I think he was on to an important point. However, I feel I have to reject the idea that awareness is based on some global brain electromagnetic field.

I have already mentioned the problem that the brain does not exist as a separate demarcated thing when it comes to electromagnetic waves. Each nerve cell will send out radio waves as an electric signal passes along its axon and activates other cells, and these radio waves are picked up by an EEG machine as 'brainwaves'. However, there is nothing in the brain like an aerial to pick up the waves.

Susan Pockett (2002), who has taken a lot of interest in electromagnetic waves and consciousness, has herself pointed out that this poses a rather serious problem. Some would argue that it does not matter but I think it does. If we want to know how we can be aware of red and fire engine together in such a way that we can talk about the fire engine being red the coming together of red and fire engine has to be a physical event (even if the awareness is not) with an output that allows us to talk about the red fire engine. If red and fire engine only come together in an electromagnetic field that does not do anything then it doesn't work. We know that the messages passing between cells at synapses are essential to talking about red fire engines. If red and fire engine come together by messages from two or more cells being sent to the same cell via synapses then suggesting they also come together in an electromagnetic blur around cells is unnecessarily complicated. It is not a question of whether or not awareness itself does anything. That is a much more complicated problem. It is that it seems absurd to suggest that sensations come together in awareness in a place unrelated to the place where the signals that we know carry information that encodes the sensations in the brain come together – in cell membranes.

Another problem with a field crossing from cell to cell is that it brings back one of the problems associated with the suggestion made by Crick (1994) and others about us being aware of signals when they are synchronised. If awareness is linked to a process with an output in the way I have just suggested, which cell is going to have the output? If a thousand cells hum together and then conclude 'yes', an electric discharge from them all would produce a violent tic, if not an epileptic fit. Putting forward the idea of bringing sensations together in some limbo 'between' cells is a bit like sawing through the branch of a tree that you are sitting on, you are left with no way back.

This sort of argument seems to dog almost all approaches to awareness that try to set it in a cluster of many cells. These cells are supposed to be passing individual messages to each other, but at the same time they are supposed to be 'in tune' in some way. It is a bit like saying that a congregation in church is both gossiping in little groups on each pew and singing a hymn in unison. It does not add up.

eight

What Waves Are Present In Cells?

So, perhaps everything so far in this story has been a flight of fancy. It is all a matter of dreaming that our awareness might be absurdly multiple on the basis of metaphysical musings. Defending metaphysics as William James's 'nothing other than an effort to think clearly' I can claim a case for writing everything so far, but what if it turns out that my suggestion is just as impossible as everything else? I have tried to persuade that it is the only avenue open that makes any sense at all. It would still be nice to see if it survives comparison with up to date information about what goes on inside individual nerve cells. If I am on the right track there ought to be some evidence for waves, or modes of oscillation, in nerve cells, which are sufficiently dominant or obvious in some context to be linked up to awareness, and ultimately to our behaviour. Cells ought to support oscillations.

This is the one place where the book goes in to physical detail and those who are really not interested can skip to the concluding paragraph of this chapter if they like. However, the ideas involved are not very complicated and I have tried to illustrate them with some every-day analogies. Reading through may at least give a general feel for what I think might be going on. It may also give some idea of the way I think we need to try to track down the slippery beast that is the listener to our story, and some of the tricks it may have in store.

Do cells support oscillations, or at least can they? At first sight the answer to this question seems to be in one sense yes and in another sense probably not the way I would like. In seeking advice on this I was advised to get it from the horse's mouth, in the person of Andrew Huxley, apparently because rather few neurobiologists actually know very much about the physics of cell membranes. Huxley had been Master of my old Cambridge College, Trinity, although when I was an undergraduate he was teaching Siân at University College London. He was kind enough to meet me for lunch. Huxley's comments to me were that the integration of electrical signals in nerve cell membranes is most certainly a matter of the behaviour of waves, variants of those named after himself and Alan Hodgkin, technically known as post-synaptic potentials or PSPs, as any good medical student knows. Moreover, in some cases these waves set up stable oscillations in voltage across the entire cell membrane. However, these are special cases and he could not see how the sort of 'chiming' wave I might be looking for could fit in, at least the way I was envisaging at the time.

Horace Barlow, another Fellow of Trinity and an authority on visual perception, also made the deflating remark that the neurophysiologists could already explain all they needed to with PSPs and were unlikely to take an interest in another speculative chiming type of wave. While having great respect for both Huxley and Barlow I was not sure that the door was totally closed and I am a great believer in their club motto *nullius in verba* – take nobody's word for it.

As I explored further, it became clear that the fine detail of how waves interact in neural membranes is still very unclear, so there may be room for hitherto unknown aspects. Michael Hausser (London and Hausser, 2005), Christof Koch and Idan Segev (2000), Steven Williams (2004) and others have shown that integration of PSPs in cells is not just a matter of adding up but, at least in some cells, is more complex and not completely explained. That could, for instance, mean some sort of pattern-based process that would provide a raison d'être for our patterned awareness, but it might point in the other direction, showing that integration occurs bit by bit in a cell, as if the cell were a branching tree of separate computer gates, not one JOIE.

The trouble is that it in my searching it was not, and is still not, entirely clear what sort of wave we need, and on several counts. It is not clear exactly what interactions in the cell might support my suggestion because the fundamental issues of physics, and perhaps metaphysics, involved are under dispute. Moreover, it is almost certain that different cells in different parts of the brain integrate their PSP waves using a combination of different optional sets of rules. Maybe it will be twenty years before it is possible to do the experiments that might test all the possible mechanisms. Maybe they cannot be tested. All that seems possible at this stage is to get some idea of what waves nerves are likely to support and see where they get us.

I was further heartened when I discovered, at our first meeting, that Steven Sevush had come to very much the same sort of conclusions as I.

Focusing on the cell membrane

When looking for waves in nerve cells that might relate to awareness it seems reasonable to assume that they will be set up either in or around the cell membrane or perhaps in attached structures oscillating in sympathy with events in the membrane. The cell membrane is where the information is.

It is probably relevant to mention at this point the suggestion made by Stuart Hameroff (1994), that consciousness happens in microtubules. Microtubules are part of the skeleton of a cell. They are a bit like long tubular tent poles that fix the membrane in place, but also have an ability to guide the movement of chemicals around the cell like air conditioning ducting. I personally find it difficult to see how detailed information can be transferred to the interstices of the microtubules so that computation can go on there. Moreover, rather as in the previous comments about electromagnetic fields, I see a serious weakness in any idea that suggests that information is processed at some new hypothetical place when we already know that processing goes on through PSPs in the membrane. At an even more basic level I would come back to the fact that consciousness, and the puzzles of consciousness are about having an input, not processing, so I am not sure that invoking quantum processing is relevant. Nevertheless, microtubules may still be important. Microtubules might have a similar relationship to the cell membrane in the generation of waves as the bridge and sound-post of a 'cello, and also the player's fingers, do to the strings. As I will come to, the effects of dissolving the cytoskeleton on the setting up of waves in cell membranes is something that has already been studied by experiment. Most of the rest of this chapter deals directly with membranes, but microtubules should not be forgotten.

Working on the basis that my search for waves should focus on the membrane, there is still a lot of scope since the membrane is a complex multi-layered structure. Electrical PSPs are the most obvious waves to consider, but they are not the only type of wave to be observed or proposed to exist in and around cell membranes. Herbert Fröhlich (1968) suggested that a high frequency electromechanical oscillation might occur in cell membranes. The existence of this wave is doubtful, but Alexander Petrov (1999) has made a detailed study of electromechanical waves of a different sort, which can be demonstrated experimentally. I will clarify what I mean by electromechanical when I come to these waves individually. Petrov's waves are a form of piezoelectricity, and probably not the only form possible in living membranes. In humans, piezoelectric effects are most clearly demonstrable in certain receptor cells in the inner ear. In single celled animals like paramecium waves of beating of hair like cilia are also associated with waves of electrical disturbance in and around the membrane.

I will try to give a practical visual idea of what these waves are like. For the interested reader I would strongly recommend Alwyn Scott's book, *Non-Linear Science* (Scott, 2003), which gives a very readable introduction to the mathematical basis for waves of the sort involved. Scott has a keen interest in the importance of waves, and non-linear waves in particular, to the mind, although his view is different from mine.

Electrical waves: Hodgkin, Huxley, action potentials, PSPs

Nerve cell membranes are well known to support electrical waves, but rather unusual ones, which is why it took the genius of Hodgkin and Huxley to work out how they might work. Hodgkin and Huxley first described the wave that carries a signal down the cell's axon, the *action potential*. This is often referred to specifically as the 'H-H' wave. However, as Huxley points out, it is a specialised form of a more general type of wave and I will use the term H-H wave for this more general phenomenon for convenience here.

In the electric currents we are most familiar with electrons flow through a solid material. The electrical waves in cells are based on currents in which charged atoms, i.e. ions, move through water. Although a nerve cell has a complicated shape it is simplest to think of it as a tube when considering H-H waves. This tube has lots of tiny pores in it, through which ions pass in or out if the pore is open. Once inside the cell the ions spread out through the fluid within the cell, the cytoplasm. Thus current does not flow along the membrane itself, which is an insulator, but rather currents flowing around the membrane create changes in voltage across the membrane that move along as the ions move. As I will come to later these patterns of change of voltage might not be considered true waves, but because at least the full blown action potential is more than just a passive electrical current it does take on certain features of a wave and can even, under special circumstances, oscillate.

What make the electrical events around a nerve cell membrane interesting and complicated are mechanisms for opening and closing the pores in the membrane at specific points, and also mechanisms for pumping ions back out of the cell (or in for other ions) to restore the ionic gradients which produce the voltage across the membrane in the first place. Interactions between these mechanisms create the full-blown action potential, which I will first describe. To use a picturesque analogy, imagine that sodium ions (potassium and calcium ions get involved too but I want to keep things very simple) are schoolchildren full of energy, dying to go to play. If allowed to they will run around anywhere. When the nerve cell is at rest they are where the teachers (ion pumps) have put them; sitting in classrooms (outside the cell) alongside a tube-like corridor (the cell). An H-H wave starts when a chemical signal at a synapse, usually triggered by a message from another cell, opens a set of doors at one point. Sodium ion children rush in to the tube and go on running about partly at random and partly to get away from each other (their electric charges repel each other) so that in theory they will eventually be evenly distributed inside and outside of the tube.

The rushing of the sodium ions into the tube is an electric current and the spreading out of the ions inside the cell is also a current. There would have been a few sodium ions in the cell before the wave started and in theory some of these may rush the other way, out of the cell, but because there are so few the net result is a current of positive ions into the cell and along the tube.

When an H-H wave is set up by no means all the available sodium ions rush in to the cell. Rapidly afterwards the doors are closed again. Only a few children ever get out of a classroom at a time. This means that even in a nerve fibre with no energy supply to pump ions out again you can set up several H-H waves one after another before the fibre finally fills up with sodium.

The electric current passing along the tube described so far is simply a pulse of ions moving down an electrical gradient. However, Hodgkin and Huxley (1952) discovered that the real wave is more complex than this. It is in a sense an active wave. Imagine that as the children bursting in to the corridor start filling it up they start knocking the door handles with their bags so that more doors open and more sodium ion children start coming in to the corridor. We now have a wave of children coming down the corridor that keeps reinforcing itself. As a result it moves faster than the simple pulse of diffusion, it is now a tidal wave. Mathematically, a tidal wave or shock wave is called a non-linear wave. In a sense it is a pulse in which the back of the pulse is constantly catching up with the front of the pulse, so it tends to be a steep wave and a wave that keeps going.

In the dendrites of neurones where signals come in at synapses the electrical waves set up are weaker than full action potentials; they are less non-linear and do not keep going indefinitely. These are the PSPs. Only if enough PSPs interact to build up a wave of sufficient strength does the cell send on an action potential down its axon. PSPs may be based on passive diffusional currents due to pores opening locally without any further opening of pores as the change in voltage passes along the dendrite. On the other hand they may resemble the full-blown H-H wave but not be powerful enough to open enough pores as they go to keep going indefinitely. Although these are theoretically still non-linear tidal waves their natural fate is to peter out if they do not meet another wave at the right point at the right time. Just as if you light a taper quickly it may burn for a short while and then go out and if you keep it in the lighter flame a bit longer it will remain alight and burn to the end, the wave obeys the same rules in both cases, it just behaves differently once it is above a certain strength. In many nerve cells whether or not an action potential is triggered tends to depend on the strength of electrical wave from the dendrites reaching a particular part of the centre of the cell called the axon hillock.

Fröhlich's wave

Going back to the idea that a wave is a sign of a force, any wave can be described in terms of the force or forces it is based on. In one sense all forces likely to be of relevance to living things are electrical forces. This is because the interactions between atoms, molecules and larger bits of matter are all based on the electrical force. Gravitational interactions inside the brain are far too tiny to be relevant and nuclear forces do not affect day-to-day events much. However, the electrical force also shows itself on a larger scale in a variety of forms that we call mechanical or elastic forces. Moreover, these mechanical versions of the electric force can be seen as having their own separate reality since they are made up of different indivisible units, which are not the photons of electromagnetic fields, but rather phonons.

Waves frequently involve exchange of energy from one form, such as the potential energy of a coiled spring, to another, such as the kinetic energy (energy of movement) of the actively uncoiling spring and its load. In some waves one force is coupled to another so that potential energy of one form, such as that of having two negative electric charges very close together, is exchanged for potential energy of another form, such as the stretching of an elastic solid which might occur when the charges move apart. This would be an electro-mechanical oscillation.

Some of the more imaginative enthusiasts for solving the mind/body problem have drawn on an idea put forward by Herbert Fröhlich (1968). Fröhlich had made a major contribution to the understanding of superconductivity and was interested in waves in which all the quanta are in the same phase of cycle, or quantum-coherent. The obvious example is a LASER: a beam of light with all the photons cycling in unison. Fröhlich used theoretical mathematical arguments to support the idea that a quantum-coherent wave might exist in a cell membrane. Many people have felt that such a wave would be able to explain awareness in a way that other waves could not. I am not quite sure why people thought this but it is something to do with the idea of all the elements being unified. Having all the units in phase does not make any difference to the fact that the wave is an indivisible pattern which can only come in units of complete pattern and which can have SAMEDI for things it passes through.

The details of Fröhlich's physics are beyond me in places but I can give a rough idea of how I understand his 1968 paper. Fröhlich noted that the voltage across the cell membrane is enormous for so thin a structure. Small changes in the distance between charges on one side of the membrane and the other could induce major changes in electrical potential energy. Such changes would also tend to distort the shape of molecules in the membrane, which would alter the elastic potential energy of the system. Fröhlich suggested that potential energy might bat back and forth between electrical and elastic, in the way that a pendulum exchanges gravitational potential energy for kinetic energy. The membrane could oscillate. Fröhlich suggested that the oscillations would have a frequency of about 100 GigaHertz.

In Fröhlich's wave the membrane oscillates by becoming thicker and thinner, like a sprung mattress or sheet of bouncy castle being bounced on. There is no particular reason to think that it will oscillate in this way on its own so the suggestion is that energy associated with H-H waves sets it going. If you like, the school corridor is lined with bouncy castle so when the sodium ion children rush in the whole thing starts wobbling like mad. It is easy enough to see that if the membrane has a resonant frequency for vibrating in this thick and thin way that a wave may be set up when the membrane is disturbed by an electrical current. What is less clear is whether or not it would be quantum-coherent. Nevertheless, there are still reasons to think that a resonant electromechanical wave could be set up in a nerve cell membrane.

Resonant oscillations tend to require an organised structure and it may not be immediately clear that a cell membrane has enough order. The underlying fabric of the membrane consists of two layers of lipid (fat) molecules back to back. The molecules in these layers can drift around freely in the two dimensions of the membrane plane. However, they can only do this is they stay the right way up – usually with their highly charged end at one or other surface. The fact that the molecules are so regimented in the third dimension makes the membrane behave rather like a crystal, but because they are free in the other dimensions it is a liquid crystal, technically a smectic mesophase (soapy intermediate state), which I mention only because it sounds good. Almost by definition crystalline structures come with modes of oscillation, although their practical significance is another matter.

There is a general rule in physics that if a structure has a way of oscillating and there is energy around then some of that energy will be 'taken up' as units of that oscillation. Even if something is heated some of that heat energy will go in to each way of oscillating the thing has (modes of oscillations in crystals create anomalies in their specific heat). The phenomenon is much more obvious when mechanical energy is put in to a system, even relatively chaotic mechanical energy, and the system resonates.

Historically, the perfect example of this is the Millennium Bridge across the River Thames. The Bridge looked perfectly stable at its opening ceremony, but as soon as people started walking on it, with random footfalls, the bridge set up a violent vibrations that made people fall over. The point about

Fröhlich's suggestion is that if a nerve cell membrane has patterns of oscillation like the bridge, and these are influenced by electrical forces, then the regular passage of H-H waves should set up such oscillations. They may be trivial but they may not.

Petrov, flexoelectricity and piezoelectricity

Whereas Fröhlich's wave remains something of a theoretical curiosity, over the last twenty or so years Alexander Petrov (1999) and others have been investigating a larger, slower type of electromechanical wave in cell membranes that quite clearly exists, certainly under laboratory conditions, and almost certainly, at least in some cells, during their normal functions.

Petrov's wave is different from Fröhlich's in that the membrane oscillates by bending. Rather than being a thick-thin wave it is a concave-convex wave. It is the usual wave of a diaphragm, or drum. To oscillate well a drum needs to be stiff rather than floppy. Like soap bubbles, large expanses of cell membrane would be expected to be floppy. However, at the very small scale of a cell the membrane may behave as if quite stiff, especially if it is anchored at regular intervals to rod-like structures in the cell skeleton. I do not know about nerve cells but I have seen rather similar shaped fibroblasts (connective tissue cells) with long branching arms look very stiff under the microscope. If they are separated from tissue and watched settling in a dish of fluid they keep their shape perfectly and if the fluid is disturbed may cartwheel about on their dendrites as if made of glass. Petrov and others have shown that both in cells and in larger pieces of artificial membrane there is enough stiffness to set up oscillations.

In Petrov's wave, bending of the membrane changes the electrical potential across it because charged parts of molecules on either side of the membrane become more tightly or more loosely packed. This is a form of piezoelectricity, which for a liquid crystal membrane is called flexoelectricity. The detail is given in Petrov's monograph, *Lyotropic State of Matter* (Petrov, 1999). Similarly, changing the electrical potential across a membrane tends to make it bend to counter the new electric forces.

This electromechanical coupling can form the basis of oscillations driven either by electrical or mechanical forces from outside. In the examples I have seen these oscillations usually follow the frequency of the stimulus rather than having a resonant frequency of their own. However, cells from the inner ear, involved in receiving sound signals, appear to show resonant piezoelectric oscillations with a frequency around 100KHz. These cells obviously have a specific reason for showing oscillations involving mechanical forces and as yet I have not seen evidence for resonant piezoelectric oscillations in other types of neurone. However, these may not be easy to measure in nerve cells in their natural environment. This brings us back to the possibility that observing a wave that is only there in nerve cells when we are aware (conscious) may not currently be feasible. The resonant piezoelectric effects William Brownell (1990) and colleagues have demonstrated in the outer hair cells of the inner ear may involve a pattern of oscillation not found in other cells and may be different from Petrov's flexoelectricity. Their membranes contain a protein called prestin that gives them a much bigger range of movement for a given change in electrical potential. Movements due to piezoelectricity in neurones should be smaller, but might be no less important. Brownell has suggested that piezoelectric effects might allow very accurate control of the timing of synaptic signalling in outer hair cells. The question arises as to whether this might be true for all neurones to a greater or lesser degree.

Flexoelectricity has been observed in other types of cell and may turn out to be fundamental to the way animate cells interact with their environment (Kumanovicz *et al.*, 2002; Zhang, 2001). Movement in protozoa may make use of it. Interactions between mechanical events, electrical potentials and the opening of ion channels are central to the way single-celled organisms move. It may not be too fanciful to suggest that the forces of the flexoelectric effect could be seen as the 'animate forces' that distinguish animals from plants and inanimate things. But that may yet prove a red herring for my quest, even if it has some useful meaning. As I shall come to later, although neurones seem about as 'animate' as any cell, they seem to have abandoned large-scale movement in favour of 'higher' things.

An electromechanical H-H wave?

Interestingly, an H-H wave in an isolated nerve fibre is associated with a mechanical wave, as originally shown by Iwasa and Tasaki (1980). The fibre shortens and then lengthens again, and may undergo further oscillations although in the isolated nerve these peter out very quickly. How this electromechanical coupling relates to Fröhlich's wave and Petrov's flexoelectric wave is not entirely clear, but these are only two of a range of possible forms of electromechanical coupling. Heimburg and Jackson (2005) have proposed a new explanation for the action potential in which coupling between electrical potential and a change in the molecular organisation and geometry of the membrane is central to the H-H wave itself. In this analysis the action potential is much more nearly energy conserving than in the original H-H explanation. As indicated below, this may be important to my search.

Which wave might give us awareness?

For several reasons, the H-H wave does not look the sort of wave to be the listener in a cell. It is a single pulse rather than a cyclical wave. (A cyclical form, in which the whole cell alternates between taking in positive ions and sending them out, does occur in certain types of neurone, but is unusual.) The usual H-H wave travels from one place to another and then disappears. The full blown action potential is the end result of the integration of incoming signals and, since it is not set up until the integrative process has occurred, would seem to be the wrong wave for 'receiving' and experiencing the incoming signals themselves. The smaller PSP waves in the dendrites would similarly seem to be the wrong waves because each PSP is the result of one signal coming in at one synapse. Many PSPs probably die out after a short distance. They do not seem to have anything to do with integration of *all* the incoming signals.

There are other features of H-H waves that make them seem unsuited to the role of listener I have proposed. The H-H wave as described by Hodgkin and Huxley is a dissipative wave, always going downhill and losing energy. The pumps have constantly to pump the ions back to where they started. It is nothing like the wave of a pendulum, or even a ripple on a pond. It is not a to and fro harmonic oscillation; it uses up work and gives none back to anything. It is also a non-linear wave, unlike the quantised waves described by Schrödinger's equation. These features pose problems if we want it to be the sort of wave that can exist as a single thing that could be our listener. The work of Heimburg and Jackson (2005) suggests that action potentials may not be quite as dissipative as thought, but this may not apply to more transient PSPs. I shall discuss their findings later, but I still see the H-H wave as unlikely to be what I am looking for in itself.

The advantage of something like the Petrov wave over the H-H wave, for my purposes, is that in principle it is both linear and energy-conserving, like the chime of a bell. It will not conserve energy absolutely, and, left alone, will be damped out quite rapidly, even in the normal living state. However, with a continuous energy input from H-H waves a Petrov-type wave could in principle keep going with a significant amplitude. The importance of a wave being linear and energy-conserving, is that it is this type of wave which comes in indivisible units or quanta, the stamp of approval that suggests that it may exist in itself in such a way that it can be aware of, or at least have access to information about, its own extent. A flexoelectric wave has the additional advantage that it can take the form of a stationary resonant wave that could potentially occupy the entire dendritic tree of a neurone, or at least a significant part of it. In contrast, each H-H wave seems to be only where it is passing through at the time.

I am not suggesting, however, that the H-H wave is not essential to the process. All the incoming information in neurones is in the form of H-H waves. *My inside story must be told in H-H waves* whichever way you think the mind works.

However, for the cell to be able to listen to the story as a whole cell, and potentially respond to the detailed patterns in the story as patterns my suggestion seems to require a more global cellular wave and perhaps a linear energy-conserving wave. It may also require the global wave to regulate the interaction of the H-H waves. This will need further discussion, but that might be the prediction. Before that I need to say something about what is known about the way H-H waves interact.

What is known about the way PSPs interact?

Whatever the nature of a 'listening wave' in a cell membrane, for my approach to make sense I think it is necessary for this wave to inhabit the domain where the electrical PSPs interact and thereby determine whether the cell fires an action potential. Morevover, to make sense this interaction has to involve a pattern that reflects the pattern of our experience. I need to say something, therefore, about what is known about the way PSPs interact.

Historically, the simplest theory to work on was called 'integrate and fire', which really implied add-up and fire. One PSP was not enough to fire the cell but a number of PSPs would add-up over a short period. (These would be 'positive' or excitatory PSPs known as EPSPs. There are also negative or inhibitory PSPs known as IPSPs that would tend to offset the adding up.) It was always clear that things would not be as simple as this. PSPs nearer the middle of the dendritic tree might add more than ones at the ends, for instance. Moreover, the short period of 'positive action' that occurs with the depolarising phase of the PSP is followed by a period when the effect is negative, in the sense that the membrane is resistant to further depolarisation while being re-polarised. That means that PSPs can have either positive or negative effects on each other according to their relative timing. However, the simple idea stood as something to test things against in the absence of enough information to build a more complicated theory.

In the last five years experiments have been devised which suggest that indeed, the nerve cell is not just a dartboard where you add up the points, with more for the middle. At least for some neurones it may be more like a poker game where a straight flush or three of a kind is what counts; a pattern.

Experiments on locust neurones suggest that the integrate and fire model really will not explain the facts. There is a neurone in the locust brain that seems to have the job of telling the locust that something is coming nearer (Koch and Segev, 2000). (A locust may or may not have a 'story of me' so the message may not be 'something coming nearer to me', just something coming nearer.) Neurones that do this sort of job have been recognised for a long time and it is clear that at least some neurones respond to patterns. The trouble is that in theory an add-up and fire neurone can respond to a pattern if the input is wired up right, so that there are only inputs for things that belong to the pattern. However, for this you really need one neurone for each possible example of the pattern, say something coming nearer from north, from south, from east and from west. If a neurone can respond to a pattern in the way that I would like, one neurone can cope with all these, but not if it is just add-up and fire. A cell that recognises a full house or a straight flush must use 'if this and this but not that, or that and that but not this' rules.

Koch and Segev's experiments on locust neurones seem to suggest that they are not just adding up. The interaction between PSPs seems in some cases more like multiplication than adding. An ace may be ten points but two aces a hundred points, not twenty. Related work by Polsky *et al.* (2004) shows that the interaction between PSPs depends on how close their starting points are and perhaps also on whether or not they are on the same dendritic branch. In these studies close PSPs more than add up but far apart ones do not.

The most interesting possibility I see in the work of people like Polsky, Koch and Segev, Hausser, and Williams is that the neurone can draw on different rules to govern the way it integrates inputs and that each neurone can use rules relevant to its job. A touch receptor at the base of an eyelash designed to make you blink at the slightest tweak might be set up just to fire if any stimulus came in. A neurone in the visual cortex used for recognising straight lines might be tuned to fire according to pattern-based rules applied with equal weight across a mass of inputs. Neurones in the prefrontal areas link-ing in to emotion and language might have complicated layers of rules so that inputs start behaving a bit like pieces on a chess board, with a significance both relating to which input they are (pawn or bishop) and how their timing relates to other inputs (behind or in front of a pawn etc.).

These ideas may be pure dreaming but my guess is that neuroscientists have secretly hoped that things are as interest-ing as this for a long time and that there has always been quite a lot of indirect evidence to suggest that they may be.

By the time this book gets read, if it ever does, things may have moved on, but I suspect towards pattern-based complex-ity. I could mention a variety of other experiments, but readers can easily find these on line. The message for 2006 seems fairly straightforward; things are getting interesting inside cells but it is still only possible to study the interaction of very few inputs in very artificial situations. Stepping back and thinking about it, if integration of information in cells involves a pattern of waves with the complexity of the sound of Mahler's sixth symphony in full swing then current experiments, at the level of 'Teacher and I: Cello Duets, Book I page 1', might be expected to miss the whole story. The wave I am looking for may only start to appear in a cell once there is a steady flow of 100,000 inputs per second in an appropriately synchronised pattern. That might be what happens when we wake up in the morning. Studying cells in the laboratory stimulated at two or three places at once may indicate rules for integration that in the aware cell simply do not apply. Nevertheless, I would not denigrate these studies in any way. They are exciting because they show that one day the full rules may be pieced together.

A two-wave hypothesis

There can be little doubt that the integration of information coming in to a neurone is based on the interactions between PSPs. Those interactions do not seem at first sight to generate a wave that would comfortably provide an indivisible observer. The coexistence of a flexoelectric wave, or something similar, in a neuronal membrane is not too implausible, but it would be nice to have some idea what the relationship between the two types of wave might be.

An idea came to my mind in about 2004 that can be illustrated by going back to the children in the corridor. Imagine that the tube-like corridor of the cell is in fact like the Millennium Bridge, it tends to set up a resonant wobble when people are running about in it. If PSP waves of sodium ion children come in to the corridor from time to time at single sites local Petrov-type wobbles may set up but die out. However, if waves of children are coming in more often and in several places the whole corridor may set up a resonant wobble.

Let us suggest two further things apply. Firstly, it is not just the number of children that makes the corridor wobble, it depends on whether the positions of the children fall into certain patterns. Maybe if children are all coming out halfway between the bridge supports of the corridor (where the cytoskeletal microtubules attach?) it is much worse than if they are coming out anywhere. In fact it may be that if children come out where the supports are they cancel out the wobbling effect of children in between. This is by no means unreasonable.

The second suggestion is that if the corridor wobbles enough it makes classroom doors come open. The neural equivalent of this is plausible. There are a number of situations in which flexing of cell membranes not only causes a voltage change but also opens up ion channels. Now you can see that if we open doors to classrooms 5, 15, 25 and 35 and not classrooms 10, 20 and 30 we may have a situation where the wobbling of the corridor opens all the other doors and children pour down to the end. Opening doors 9, 11, 19, 21, and 29 might let more children in to the corridor in the first place but might get nowhere because the corridor fails to wobble.

An important condition for such a way of looking at integration of incoming signals in a neurone is that the incoming signals would need to be synchronised. As I mentioned previously, after each H-H wave there is period when ions are being pumped back out and the local electrical change is reversed. It is as if after a batch of children comes out of class there is a short period in which a teacher rounds up both these and any other loose children and puts them back in class. So the cell can only be expected to respond in a subtle way to complex patterns of ions coming in at different synapses if timing is perfect.

The evidence is that timing can be very precise in patterns of nerve firing. A need for synchronisation makes sense of the fact that Crick (1994) and others have suggested that what we are aware of is based on synchronised impulses firing about forty times a second. The idea that this synchronisation in itself would make all the signals available as a pattern while they are still travelling down separate nerves is absurd, as I have said, but it makes a lot of sense to say that without synchronisation it would be pretty impossible for us to be aware of patterns wherever the signals are being brought together.

The basic idea is that the interaction of PSPs is regulated or moderated by an electromechanical oscillation in the membrane rather in the way that an escapement regulates the driving of the hands of a clock by a spring. The PSPs and the electromechanical waves would be coupled. Energy would flow from PSP to the electromechanical wave. The electromechanical wave would in turn modulate the interactions between PSPs. Technically speaking, the effect of the electromechanical wave on the interacting PSPs would show itself as time and space dependent variations in the capacitance and conductance terms of the H-H equation governing the PSPs. In practical terms the output from our cells would be guided by aware waves.

It may be too early to judge the significance of the new theoretical model for the H-H wave proposed by Heimburg and Jackson (2005). However, it may provide some support for my approach. The suggestion is that coupling of electrical and mechanical forces is an integral part of the H-H wave. For the full action potential this coupling leads to the formation of an energy-conserving form of wave called a soliton. This implies that the capacitance terms in the original H-H analysis may vary more than previously supposed. PSPs in dendrites would probably not take this soliton form; solitons tend not to interact with each other and the different conditions in the dendrites would probably not support a soliton. However, it seems reasonable to presume that electromechanical coupling occurs in the dendrites as well. This is precisely what I need to 'kick' an aware resonant electromechanical wave into action, or perhaps to provide the constant patter of footfalls needed to get the Millennium Bridge wobbling.

Summary

I will conclude this section with a brief summary, partly for those who may have chosen to skip the details. I think there is reason to believe that brain cells may support oscillations of the sort that seem to be needed for a cell to be a listener in the way I am suggesting. The oscillation, if it exists, is likely to be some sort of electromechanical or piezoelectric 'buzzing' in the cell membrane. I cannot prove it is there but at least there may be some point in trying to go out and find it. This oscillation must coexist with, and in a sense listen to, the electrical waves generated by signals coming in at synapses. The precise relationship between these two phenomena is unclear at this stage but might be deducible from both theoretical and practical work.

nine

Cogito Ergo Sum

Having thought about various different sorts of waves in cell membranes and before I leave physics behind, I would like to come back to the question of what exists in such a way that it can be aware. For centuries we have worked on the basis that things exist. Various people like Plato, Spinoza and Berkeley have cautioned that we may not quite know what we mean by existence, but the general idea has survived. Since the late nineteenth century, physics has, however, made things more difficult.

Physics has, firstly, highlighted just how unsatisfactory things like hardness, weight and opacity are as signs of reality. 'Stuff' is not quite that simple. Secondly, and more seriously, physics might seem to lend weight to Berkeley's view that we can only talk of what we experience, not of what *is* out there (Berman, 2001). In standard quantum theory existence seems to be reduced to observables, only appearances, not enduring realities. David Bohm (Bohm and Hiley, 1995) tried to re-formulate the theory so that there could be beables (think be-ables), that existed somewhere in space, even when not observed. However, Bohm's approach has its own problems. His major insight may have been to show that it may be possible to find a better way of looking at existence than the standard interpretation of quantum theory, but it may need a more radical approach than even he could produce. We may be able to get back to the idea that things *are* things but we may need to give up some old ways of thinking.

Some might say that we should forget about the idea that there *are* things, that existence is a valid concept. The problem with this is not so much one of objects as of subjects.

The existence of objects can be open to doubt in a variety of ways. A greengage tree may seem to be a thing but there are uncertainties. Its trunk may be more firmly attached to the ground than to its leaves. Tracing the roots you may find that the tree is merely a sucker of another tree. If you cut the bark and sap flows, is it part of the tree, and when does it stop being part of the tree? I know a greengage tree that was in the same place from 1975 to 1989 and was also there in 1995 but I cannot be absolutely sure it exists now, or even that it was the same tree throughout the period I knew it. These may seem like quibbles but physics tells us that at the fundamental level they loom very large. If existence was purely an issue about objects it might be as well to say that it is no more than a concept of convenience and often a matter of opinion. In short, whether the earth is really a planet or really two half planets does not matter.

However, for observing subjects it does seem to matter. When Descartes said *'cogito, ergo sum'* he was saying the he must exist *as a subject*. There would seem to be no room for half subjects.

One way of looking at things is that in any version of the universe there is only ever one subject. For me it is me. For you it would be you, but that would not be my universe so would not really exist. It is, however, more interesting to consider the universe as containing lots of subjects which, at least for themselves, have to exist. Moreover they have to have some sort of intrinsic existence as separate things, which seems to mean that they have some sort of defined locus or domain such that they have access to whatever manifestations of the universe 'belong' to that locus or domain.

At present objects are broken down into components that behave as waves (which may also appear as 'particles') of various sorts. Subjects are treated as massive conglomerates of matter that lie beyond the laws governing objects. This is about as unsatisfactory as you can get. It is almost as if we are looking at physics down the wrong end of the telescope. It is subjects that we need to be separate elements; our experience tends to deal with objects all together. As will become clear later, I have a suspicion this may be precisely where we are going wrong.

What would seem to be the happiest way out of the problem is the possibility that being an elemental object corresponds in some way to being an elemental subject. We should then look for things whose behaviour as objects would make them candidates for the status of our sort of subject.

What we seem to be looking for are things that are indivisible, or non-decomposable, at least in some respect, that can have a single set of instructions about how they relate to the rest of the universe and which seem to have some sort of continued reality through time, even if a very brief time, rather than just being a transitory appearance. Modern physics readily provides us with objects of this sort, in the form of whatever things are described by a wave function; quantised wave modes. Although physicists may be coy about this the fact that these modes always come as a whole number of indivisible units of complete wave seems as good an indication that they have an intrinsic existence as any.

Quantum physicists have two sorts of reservations about saying things exist. They like to avoid saying what is really there except when you observe it, and for good reasons, but this is more an issue about where the thing is or how much momentum it has rather than whether or not it exists. They would also probably object to the idea that a wave function exists, since a wave function would seem to be a set of instructions for predicting what one might find rather than a thing in itself. I will come back to this issue later but for the moment I will stick to the wording in the last paragraph 'whatever thing is described by the wave function'. I suspect that most physicists would be reasonably happy to allow such a thing to exist, as long as our knowledge of it obeyed Heisenberg's Uncertainty Principle.

This seems to provide us with a basis for some of the thoughts I have already introduced. Since the chime of a bell is a mode of oscillation that can be described in terms of quanta it seems legitimate to say that a chiming bell involves a separate indivisible thing that really does exist as an object, even when we are too distant to hear it, in the same way that an individual electron exists as an object or 'observable'. If we then turn things around and consider such an indivisible thing as a subject, or as an observer, the result seems rather satisfactory. The bell would be aware of all the perturbations that might affect its mode of oscillation through a single indivisible relationship, because that is the way relationships with waves work.

The potential problem with allowing something as big as a bell to be a single indivisible existing 'thing' because we can find a mathematical formula for a wave that might occupy it is that one might argue that any large object might qualify. As long as one could find some sort of formula that would describe some sort of wavelike activity one might have grounds for being a subject. One might say the Wall Street could be a subject because the stock values go up and down in a wavelike way. Rhythmic cycling of electrical currents in a computer might make it a subject. However, a status as subject would only be applicable in relation to those things the subject could observe, or interact with. This would be similar to the status of an object or observable, which is only such in terms of those observers that can, directly or indirectly, observe it.

Thus, the wavelike fluctuation of stocks and shares would only be an observer for those things that interact with those wavelike fluctuations in an indivisible way. This makes things easy because nothing very much does that. Nor does anything interact with the fluctuations of electrical activity in a computer very much. The chime of a bell would only make the bell aware of its mechanical constitution and environment. A piezoelectric mode would make a cell membrane aware of the myriads of electrical inputs that we know carry the information that fills our 'consciousness'.

This means that it may not be too difficult to separate 'bona fide' potential subjects, in a sense that we are seriously interested in, associated with quantised waves, from things that bear only a passing resemblance. I am reasonably confident that the traffic of information in a computer or in a brain as a whole is sufficiently unlike a quantised wave to be disqualified. When a 'functionalist' says that a brain or computer can be aware because it is an information processing system I do not think they have the justification for treating the system as a single subject we might have for a bell or a cell membrane.

If a light beam is reduced to a minimum we find we cannot break up its pattern. The information from one of Young's slits is always bound together with the information from the other slit. In contrast in a 'system' such as a computer in which bits of information are trafficking around this is not the case in any sense that we have evidence for. If you replaced the semiconductors in the computer with a team of laid back Mexicans who can be relied on, eventually, to take, process and deliver messages in the right order, but you never quite know which day of the week, the functionalist should say the system would still be aware. Apart from this seeming barmy it illustrates the fact that in a computer, or a nerve net, separate events occur separately, rather than beings facets of one indivisible wave-based event. The idea of a 'system' adds nothing to the causal process, it is an optional convenience for a quick description and nothing more. Force fields like light beams and bell chimes do not work this way and without the concept of the field as a whole predictions about what will occur will be wrong, even if the difference may be slight. These seem sounder grounds for 'existing'.

Leibniz had a clear view. There are elements, which are subjects, and these progress in harmony with everything they have access to through a single act of perception. In a sense there are no individual objects. Our studies of the behaviour of a single quantum of energy are, in this sense, studies of its behaviour as a subject, not as an object. There is no half way house between an element and the universe that could exist as a system. System is the name of many subjects progressing in harmony with the universe. I suspect he was right.

One way of looking at the problem is to ask 'does information bring a rich history with it'? Each bit of information in a computer or a neural net brings nothing with it other than the one 'bit' that it is. In contrast a wave passing through tunnels or an x-ray beam passing through a crystal brings with it the richness of its history. When someone speaks to us we only receive the words, not the experience that prompted the person to produce the words; we are not telepathic. The same should apply to one cell communicating with another in the brain. Only within a wave does physics allow richness.

What I am still uncertain about is exactly how limited the rules are which would define the type of wave needed to exist as an aware individual. Linear waves that conserve energy, to a first approximation at least, like a beam of light or the chime of a bell, seem to have the best claim to being real. The difficulty comes when the wave is either non-linear or non-energy conserving. Alwyn Scott (2003) uses the example of a candle flame as something which seems to be an individual 'thing' which is a non-linear non-energy-conserving wave. But I am not sure it is a thing in the sense that it might be aware of its domain. In many ways it seems much more like a pattern of things passing through rather than a thing in itself. The H-H wave is similar in at least some respects. What I cannot be sure of is whether such a wave fails to pass the test of individual existence because it is non-energy-conserving, because it is non-linear or because of either or both. These are things I still need to try and find an answer to but remarkably few people are ready to give an opinion! Maybe that is because there is no final authority we can turn to.

Spinoza and the nature of existence

There are certain things about waves which draw me into metaphysics a little further. I do not mean metaphysics in a mystical sense. I am the sort of person who sees that Buddhism may be a rather nice way of looking at things but is not a Buddhist. My empathy is with David Hume (1999), in just wanting to find a view of our relationship to the world that ultimately fits with stringent logic.

The more I think about awareness, the more I come to believe that our problem is not that we do not understand how the brain works, it is that physics has got stuck in a way of thinking that prevents us from making sense of what we know. Physics sets out to explain the way things are. In Newton's time it seemed to do fairly well. In comparison today's physics explains nothing. It is simply a set of recipes for predicting what measurements you will get. Where Newton had some underlying concepts that 'made sense' which guided him through his road to his discoveries, modern physics makes no sense, or at best very dislocated sense. What I find difficult to judge is whether this is because there is something missing from the physical theories or whether physics does actually make sense but nobody is looking at it from the right angle. Maybe physicists are really rather conservative and cannot quite make the leap to a new reality which their physics spells out quite clearly.

The problem all boils down to what we actually mean by existence or reality. Philosophical views on this have famously swung back and forth with Plato, Locke, Berkeley, Hume, Kant, Schopenhauer, Wittgenstein and a dozen more. Yet in a sense we can agree with all these people because to a great extent they all express the intuitive human view; just from different standpoints, particularly in relations to use of words. (I find it hard to agree with Wittgenstein, but I do not deny that his standpoint can be taken.)

In Spanish there are two verbs to be; *ser*, to be in essence, and *estar*, to be in a place. They are the same in Portuguese, the family language of Benedict de Spinoza (Scruton, 2001). In trying to understand quantum physics I have repeatedly asked myself whether the bits of it that seem so odd, the wave-particle dichotomy, the 'measurement problem' and 'collapse of the wave function' are only odd because we confuse different senses of being. Although the parallel is not exact I might suggest that people may have confused *ser* with *estar*. I was intrigued to be told by my colleague Maria Leandro that in Portuguese schools the wave-particle dichotomy is taught in precisely this way. The wave is the *ser* and should be seen differently from the particle, which is the *estar*. However, I wonder whether the full significance of this difference has sunk in. If Spinoza were alive today I would be interested in his view.

Spinoza tried to solve the mystery of what created the universe by stating that there is something, which could either be called Nature or God, and which included all aspects of the universe we can know, and more, which exists *because it is in its nature to exist*. Nature/God is what existence is and so there is no possibility of it not existing. Quite what one makes of this is difficult to be sure of but it seems to have advantages over the idea that we have to say there is a God that created us, because you would then need a GodGod to create God and so on.

One of the ways that Spinoza's description of Nature is translated is as 'cause of itself'. That is, because its nature is existence, it requires nothing else to bring it in to existence, it generates its own existence. There is an interesting aspect of waves that seems to echo this. The fundamental wave patterns that modern physics puts at the heart of things are complex relatives of the most familiar wave, a simple harmonic oscillation, such as a sine wave. It is the smooth up and down, up and down, curve that underlies vibrations in strings, ripples on a pond and many other familiar waves.

If one was to ask what is the cause of a harmonic oscillation one might be close to asking what is the force behind it. In Newton's Laws of Motion, force is closely linked to acceleration. Acceleration is technically the second differential in time of something; in ordinary language the speed at which the speed of the thing is changing. For a wave it shows up not as the slope but as the way the slope changes, how much it curves. The curve of a harmonic oscillation changes back and forth from one direction to another. In fact the pattern of the curve of a harmonic oscillation is a harmonic oscillation of the same type. So perhaps a harmonic oscillation is the closest one might get mathematically to 'cause of itself'? After all, if we ask why a clock's pendulum is swinging, a perfectly valid answer would be because it was swinging just before. Swinging causes swinging.

Fanciful maybe, but there is another aspect to the fundamental waves of modern physics which might have interested Spinoza. If one asks what are they waves of, they are not waves of movement of matter, nor of some mystical ether, nor are they waves of force, they are *waves of likeliness that so-and-so will appear to be so*. What a fundamental wave is to a physicist is the likelihood of finding something like a certain amount of momentum or energy at a certain point in space-time. It is a set of instructions.

Spinoza said that Nature exists because it must exist, it is *certain* to exist. This raises the suggestion that likelihood, or probability, is at the root of existence. So if we want something the nature of which is about being likely to be so and which seems to have the same form as whatever might cause it, the waves of fundamental physics seem rather well suited. In fact, as far as I can see recent concepts of 'quantum cosmology' produced by people like Steven Hawking seem to suggest precisely that.

A potential problem with Spinoza's view is that he says that the one thing that must be so is Nature, the whole universe, yet in a sense the one thing that has to be so for me is me. If I ask what is the probability that there is at least one thing that is receiving a story, and that it is in my head and witnessing this book being written, I think I can reasonably suggest that the answer is 1; certainty. If I ask why am I me and why am I aware I could be said to be asking what is the likelihood, given every possibility that might ever be, that I am me and aware. The answer is that the likelihood must be absolute certainty for me to ask the question.

This problem seems to arise from the difference between what is likely to be so and what is likely to appear to be so. The latter implies a subject, in effect it implies me (or for you, you). My thought is that we have to accept that there is a subtle relationship between what is so and what appears to be so, which modern physics tells us about in some detail, and that we need to consider that there are two sorts of reality, or existence. There seems to be an 'underlying existence' (tending towards *ser*) which comes as packages of instructions about likely appearances and also an 'apparent reality' (tending towards *estar*) of the appearances themselves.

Neither of these two types of reality can stand alone as 'the true reality'. In one sense underlying reality is the primary reality because the morning after a wild party in which everyone ingests intoxicating substances there is eventually a sense that something was going on out there based on reliable sets of rules which brings us all to the same view of a present world in the end even if we remember nothing about what happened. On the other hand apparent reality is primary in that at least we have immediate access to it, we have to believe in it, whereas we may discover that our idea of the underlying rules is flawed in some way.

The key significance of distinguishing these two types of existence is that it is simply inappropriate to consider the detail of underlying reality in terms of apparent reality; they are based on different types of fabric. David Bohm tried to find a way to attach an appearance to the packages of instructions at all times but the result, however mathematically sound, did not have the sort of appearance we are familiar with. Particles had to leap about in a way that rather spoils the comfort of knowing 'they are always somewhere'. My guess is that the conventional view of quantum theory, which states that the packets of instructions do not give rise to any fixed 'unobserved appearance' between actual appearances is ultimately the right approach. The things that have 'existence throughout' neither jump about nor spread out or go fuzzy or anything like that. They are fixed *sets of instructions* that get 'read' when an appearance is experienced. These sets of instructions usually come in a few standard forms, like electron, or elastic force field, because they have a finite likelihood of being so.

An important implication of this way of looking at things is that it avoids the idea that reality involves the strange two-step process described by von Neumann, demarcated by the mysterious event called wave function collapse (see Penrose, 1994). There is nothing that spreads out like a wave and then 'collapses' down to a particle in one place, other than the uncertainty of the observer about what will be the next appearance. I am puzzled by the suggestion of Roger Penrose that one should look for ways in which wave functions might collapse in a sense other than an observer's uncertainty disappearing.

The existence of this built in uncertainty at the quantum level does, nevertheless, need some sort of explanation. It seems that every time there is an interaction in the universe, in which energy shifts from one wave mode to another, a random factor is injected into the world which together with Schrödinger's equation, will determine what the next interaction, that can give rise to a certain appearance, will be. This should come as no surprise if every newly born quantum will need to have a value for, for instance, phase angle. Since there would appear to be an infinite number of possible angle, spanning 360°, of equal likelihood, the choice really has to be random.

In order to find out about how something like an electron is behaving it is necessary for something, like a photon of light, to interact with the electron in such a way that it can tell you something about the electron. This is called making a measurement. It does not require anyone to look to see what the photon can tell us, so it has nothing, as yet, to do with observation or appearance. However, when the photon interacts with the electron it alters something about the electron. As an example it may seem to produce a local, and unpredictable, shift in the phase of cycle of the wave that determines where the electron may appear to be later. This gives the impression of stopping the electron wave from interfering with itself, supposedly one sign that the 'wave function has collapsed'. However, as far as I can see the photon does not stop the interference effect, it simply skews it in an unpredictable way. Nor does it reduce uncertainty. The interference effect is normally observed by sending thousands of electrons along a path, like in Young's double slit experiment, and looking for an 'interference pattern' at the end of the path. This interference pattern relies on the phase of cycle of the electron waves all following the same rules. If there are unpredictably different changes in phase for each electron due to measuring photons the interference pattern becomes a smooth smeared out band, which is what it would have looked like if there were no interference. Thus interference is still going on but you can no longer demonstrate it.

The importance of the measurement effect is that you cannot observe without measuring. The measurement tends to get incorporated into the concept of the 'effect of observation'. However, as Richard Feynman pointed out, we have to assume that the measurement effect happens whether or not anybody observes. There is no effect of *observation* on the thing being observed.

When an observation is made something quite different occurs *just for the observer*. Before the observation the wave function is the best set of instructions for predicting where you will find, for instance, your electron. Once you have made the observation you know where the electron is, so the uncertainty has 'collapsed'. Mathematically this corresponds to a curious thing called Dirac's delta function, which makes the probability of all possible situations 0 except the one situation that is observed, which has a probability of 1. However, nothing more 'happens' from a third party point of view than the effect already involved in the measurement. Thus the collapse associated with observing is not something that happens 'out there' it is a collapse of a property of the observer; uncertainty about the thing being observed. Collapse not associated with observing makes no sense.

These arguments make me puzzled when I hear people talking as if consciousness is what happens when wave functions collapse. It is certainly true that awareness is all about uncertainty collapsing but from the outside perspective there is no special 'collapsing' going on inside someone else's brain. So we cannot have a theory of consciousness that applies to other people that invokes wave function collapse. Quantised waves are just going about their business as usual. People might say that because quantum waves involve uncertainty there has to be some point when the universe makes up its mind where things are going to be. But this is not what quantum theory says. The uncertainty that the wave describes is not the universe's uncertainty but the observer's uncertainty about the universe. In a sense the universe always knows exactly what is going on.

Except that there is nothing there to know anything and in a sense this means that nothing *is* going on because going on is what appearances do, underlying sets of rules do not go on. This may sound difficult to follow and it is. It is probably what made Whitehead try to say that nothing exists except (i.e. nothing in between) occasions of experience. I prefer a slightly different way of looking at it.

So the thing that quantum physics does not allow us to do is to pretend we can catch a wave function half way between two observations and think we can know 'what is going on". Despite David Bohm's attempts to overcome this it seems that this is not so much impossible as *meaningless*. One way of justifying why it should be meaningless is to say that universe is built up of sets of instructions (wave functions) each of which has a particular domain of time and space and describes a complete 'history' in that domain which is indivisible. The instructions are not for an unfolding progression of states but for a total history, delivered *en bloc*. These histories interact at junctures where energy is transferred from one form to another and these junctures can be points of observation but there is no reality in between these junctures other than the instructions.

As a crude analogy one could imagine the universe as built like a shawl made out of sequins. Every sequin is joined to other sequins at certain points. You can fold (divide up) the cloth at any of these junctures but each sequin is a complete, rigid unit. Nothing 'goes on' inside the bounds of each sequin. As far as I can judge, this concept of quantum histories being indivisible *faits accomplis* is fairly much the approach taken by Richard Feynman and as such very orthodox, although Feynman tended to disparage 'philosophy'.

Feynman's concept of a history includes the idea that a history is rich, in the sense that it arises from a 'sum over all possible histories' within the domain. What appears to have been a photon passing along a single path is in fact the result of a set of instructions for every possible path for the photon creating through interference one most probable path. If the photon is an x-ray passing through a crystal it will encounter a richness of paths reflected in the interference effect that can be made apparent by passing many photons through the crystal to a photographic plate. This is where the quantum history as subject begins to make sense. If quantum histories as objects are indivisible then it is reasonable to think that quantum histories as subjects are the same. This would mean that rather than awareness being some sort of revelation of the state of affairs 'at the point of wave function collapse' it would be an account of the entire 'bundle of possible histories' for the quantised field involved. Although it would seem to be 'now' it would in fact represent a domain of space and time, which it would seem to need to in order to be rich.

This brings me to an issue which seems to be a concern for many people trying to find a metaphysical basis for modern physics: what Steven Sevush has called the need for 'punctuation' in the universe. There is a sense that the universe cannot just be one continuous fabric of existence. In particular there seems to be a need for points in time when things 'happen' so that they can act as commas or full stops. For many people these points are points of wave function collapse but as I have indicated there are some problems with knowing what this means. Whitehead suggested that these 'happenings' or occasions were what the universe really consisted of (Whitehead, 1978). In my sequin analogy the punctuation comes at the beginning and the end of a quantum history. However, it is not immediately clear what these beginnings and ends are.

For an electron one might think that its beginning and end might be several billion years apart, perhaps from a second after the big bang to the end of time. This might not be so and I have a feeling that any discussion of the awareness that an electron might have is going to be pretty hypothetical. Nevertheless, my first guess would be to allocate one experience to this entire history, making it irrelevant to our search for our own experiences. I do not see this as a problem because in practice our familiar world is built out of the histories of much more evanescent quanta, the boson based force fields of light, sound and related mechanical forces. Defining the beginning and end of a history for a photon or a phonon would seem relatively easy. This makes me think that the punctuation we are looking for in our everyday world comes from the birth and death of bosons. Unlike wave function collapse these are objective events. Nevertheless there is a link.

Whenever a 'measurement' is made bosons are created and/or destroyed, regardless of whether somebody observes the measurement. Within the brain of an observer countless bosons are created and destroyed in the process of getting information from the eyes or ears to wherever in the brain awareness occurs. I am suggesting that the state of awareness also involves a 'subject' boson-based field coming into and going out of existence. My attempts at pinning down such a field, perhaps based on piezoelectricity, are so far still tentative but I can conceive that such a field might last for the duration between full-blown action potentials in a cell. The field would take the form of a resonance during the period in which local PSP waves evolve in the cell's dendrites. When a full self-sustaining action potential occurs the resonance might be transiently abolished, as when a pianist momentarily lifts the sustaining pedal at the end of each bar of Chopin's Grande Valse Brilliante, Opus 18, no.1.

I may be wrong, but I have a strong feeling that the problem with modern physics is not that there is anything major missing from it but that its full metaphysical implications have to be accepted at face value. As a result people think that it predicts nonsensical things that it does not, or that awareness has to be outside physics. I think we will need to adjust our idea of the relationship between existence and appearance, but in a way that is actually very close to the way we treat the world. Dirac's delta function describes something very familiar. During a game of cards the likeliness of who has the king of diamonds succumbs to the delta function once that card is played. As David Hume (1999) pointed out, probabilities play a major part in the way our brains work. The way we see things may be less a matter of working out what patterns of things are out there than working out what the likelihood is of each and all of the possible patterns we can conceive of being out there. Is it a bird?... is it a plane?... Likelihoods shift as we cast our eyes around, and sometimes jerk backwards when we realise we have seen an illusion. I would not be surprised if in a hundred years time that something like the idea that the basis of existence is the likeliness that so and so will appear to be so may be just as much accepted common sense as Newton's gravitational force was for a century or two.

Having taken physics and metaphysics as far as I can at this point I find myself with two conclusions. Firstly, there seem to me to be inescapable reasons for thinking that if awareness has a physical basis it resides in each cell separately. Secondly, the basis for awareness might fit with a transient piezoelectric oscillation with a frequency in the order of 10MHz, which would have access to the rich electrical perturbations around synapses, but this remains purely conjectural. What at least may be fair to say is that if William James were alive today he would not necessarily conclude that rich awareness is physically impossible.

Part III:
If It Were So

ten

Thinking Is More Than Awareness

While a theory is being constructed, its incomplete state of development often prevents its detailed experimental consequences from being assessed. Nevertheless [we] must make choices... Some ... are dictated by internal logical consistency... but ... some ... are founded on an aesthetic sense...

Brian Greene (on superstring theory), *The Elegant Universe*, 2000

In discussing consciousness some people appear to use the word to mean awareness and some to mean thinking. The confusion leads to all sorts of problems. In early discussions about the idea of 'consciousness' belonging to cells individually I came across a number of people who made comments like 'How could a single cell make all the necessary decisions?' or 'Why wouldn't there be conflict between the actions of different cells? Many times I have needed to point out that I am not in the least suggesting that each cell can *think* separately. Thinking is a rolling narrative of sensations followed by actions which become themselves part of the next instant's sensations which then are followed by more actions and so on. It is a series of events that must involve a net, or at least a chain of many, many cells. What I am suggesting is that each of these many cells has an enormously rich input about what is going on, and nothing in the brain has a richer input, but in terms of output, each cell is restricted to contributing only one in ten billion of the signals which determine the progress of thought. If we make an important decision every minute and one cell tips the balance of that decision the average brain cell will have this honour once every thirty thousand years. It is likely that many brain cells never determine an act throughout their entire life.

This is no more than what we know based on fact. Neurones have wonderfully rich inputs and only one output. The output may be sent to a thousand places but it, and the places it reaches, are always the same. As I indicated before, neurones can be viewed as like couch potatoes watching a video with a one button infra-red handset in their palm. Imagine the following scenario in the not too distant future. An interactive video is playing on a TV channel being watched by ten billion five year olds lying on their sofas. The viewers are told to press their buttons when they want something exciting to happen. The video is truly interactive in that there are a variety of options for the story and these are chosen on the basis of how many signals come back from the viewers at any one point in the story. The catch is that the children do not realise that anybody else has a handset as well, so each child at the end rushes out to say 'mum, I made this video really exciting by pressing my handset, I did it all myself'.

Again, if you take away the issue of where awareness is, this scenario is simply a picture of the scientific facts about the brain. The progress of the story is governed by the overall effect of lots of cells firing when they receive certain patterns of input. Moreover, one thing we can be sure about is that neurones have no way of having access to information about whether or not their output has had an effect on what their next input is. If they are aware at all, they will not be aware that their contribution is trivial, just like the children.

So if individual cells are aware there is no possibility of 'being at odds with the rest of the mind' because nothing is aware of any 'other' source of action. There will be competing and conflicting strands of the story, sights and sounds, likes and dislikes, jostling for dominance, but the cell will never have a story about itself that lets it see itself as at odds with other cells.

Perhaps one way of putting my argument is to borrow the approach of those that say that awareness is not something that needs some mysterious explanation, it is just a reflection of function. If so, we need to contrast the functions of awareness and of thinking. The function that matches awareness seems to belong to some sort of unit having lots of information coming in, including an account of actions. These actions appear to be based on the information coming in, in that they can include talking about the information coming in, and yet these actions often appear to be determined by other functional units that get to the information before it gets into the story we can talk about. Actions are part of the more global function of thinking.

With my misappropriated functionalist hat on, I can say it all makes sense; the functional unit we are looking for for awareness is a neurone. The more usual 'functionalist' view, that awareness is a reflection of the function of a net of cells computing away makes no sense; it fails on its own terms. Our awareness is not of computations, we find ourselves as if at the wine tasting, not in the middle of the vineyard amongst pickers, grape treaders, barrel makers and bottlers.

The more you think about it, the idea that awareness is something that goes with signals between cells makes no sense. Signals between cells are designed to get ready for experiencing, to get messages together; they cannot also be the experiencing. The best that a signal between cells can be is the last step in the process of sorting information for experiencing. The experiencing can only occur at a juncture, a true physical JOIE.

Oscillations of electrical signalling activity between cells, which one might describe as waves, could, and do, occur, such as the rhythmical patterns seen on an EEG. We might start worrying about whether or not this wave is real and so can be the site of experience. But that would not give us the experience we have; it would be a wave of getting ready for experience. It would have no output as a wave, just the individual outputs from the cells getting signals. We do not need or want to experience the work at the vineyard. There is no point. We want to taste the wine when it is ready. Put another way, if a wave of signals had a function it would have to be a different function from the one we know the signals do have – each one a message from one cell to another. It makes no sense to suggest signals to have two functions; it would produce a conflict. I find myself arguing this as if I were trying to counter some mistaken dogma that has taken over the scientific world. Yet I am only expressing what is in the textbooks; the 1968 edition of

Mountcastle's *Medical Physiology* (Mountcastle, 1968) I first learnt my neurology from says almost exactly what I have just said.

So, although my suggestion seems to throw everything out of kilter in terms of the way we are used to looking at things, if we put it alongside the scientific facts it marries up rather better than what we thought was how things worked. The next question is how the view of one story with many listeners might affect our understanding of the way the story is built up, of how thinking works as a whole.

Hierarchies, holograms and fractals

While it true to say that most of the neurones in the brain function in much the same way, with input dendrites and output axons, it is hard to believe that the brain is simply a colony of ten billion equivalent cells chatting to each other without any sort of hierarchy or division of labour. Division of labour is relatively easy to identify in the sense that we know a lot about which cells deal with, for instance, vision rather than hearing. Identifying hierarchies is more tricky. Asking which cells are in charge produces no clear answer. All cells are both controllers and controlled.

What we do know about are the major connection pathways that allow cells to send each other messages. Messages tend to come in to the brain stem. The brain stem exchanges messages with the thalamus and other central clusters of cells. The thalamus exchanges messages with the cortex through thalamo-cortical pathways. The left and right cortices exchange messages through the corpus callosum. Other interactions occur within the cortex and it has been suggested that exchange between the prefrontal and parietal lobes is particularly important for maintaining a narrative of self in an ever-changing context.

Perhaps what we want to know for each of these exchange pathways is which cells are asking the questions and which giving the answers. But it seems that cells at both ends are doing much the same sort of thing. Various people have tried to define some general set of rules for interaction between cells that might make sense of this but there seems to be no clear consensus. There is a very large artificial intelligence literature that aims to provide theories of how things work. However, despite pages of mathematical formulae I find it hard to discover any clear simple ideas derived from such theories that can be applied to real brains.

There can be no doubt that an understanding of how information can be manipulated is a hugely powerful thing. You can build space ships and mobile phones that do remarkable things. But when I fight my way through theoretical models of information processing in the brain and feel I grasp their meaning I get the impression of nothing more than a very complicated description of the obvious possibilities. Moreover, neither does there seem to be any way of confirming the validity of these theories beyond very simple actions like making still things appear to stay still when you move your eyes or body. In particular, these theories do not seem to be any help in understanding awareness because they do not put awareness in any particular place. It is supposed to arise out of everywhere at once.

I am very happy to believe that there are people with insights into information processing which I have not spotted and which I may never be able to master. However, in other fields where very few people are in the know, like quantum physics, one can still appreciate the new concepts and how one might derive these from the detailed mathematics if one had the time to go through all the layers. Nevertheless, there are a few accounts of information processing that do seem to have some relevance to the way awareness in single cells might be built in to a functioning brain.

Karl Pribram (1991) and a number of colleagues have taken a particular interest in the idea that the brain may work somewhat like a hologram. A hologram takes a patterned array, such as a picture, and converts into another array in which every point of the new array tells you about an aspect of the entire original pattern. The detailed way in which holograms work is not really relevant here. What matters is the idea that you can store and use information about an image in the form of a set of relationships rather than an array of dots. There are lots of variations on this idea. Some use a set of strict mathematical rules known as Fourier analysis, but there are looser versions of the idea, and a nerve net would be expected to approximate to, rather than fit exactly with any set of mathematical rules. A very simple example would be a description of a tablecloth which was not based on what colour each pixel is in turn but on its patterns: blue and white check; check squares one inch across; roses in the white squares; five petals to each rose; petals coloured pink. This sort of approach obviously has attractions in the context of my earlier comments about patterns and the binding problem, It will also be relevant in the next chapter when discussing the language of awareness.

Pribram was interested in holographic processes because of some interesting observations about how the brain analyses visual images. Rather than just recognising 'blobs' parts of the brain seem to recognise things like stripyness. I suspect there was also an interest in the fact that processing of information in the brain seems to get spread out much more than you might expect from the sort of processing we see in an ordinary computer. Any particular sensory input seems to be sent out quite widely for processing and, moreover, any one bit of brain seems to get involved in processing quite a variety of inputs. There has been a long debate about how much the handling of information is 'distributed' about the brain, but there seems to be evidence that it is to a significant extent.

My impression is that Pribram and others may have considered that what happens when a sensory input comes is roughly as follows. All the different elements of the input are sent to lots of different neurones, and that in these neurones the information is present as a 'holographic transform' which, if you print it out on paper as if it were itself an image, tends to look like a strange pattern on a radar screen rather than a picture. This gobbledygook image is then analysed, perhaps in the context of memory, and sent back to some integrating centre. An attraction of considering the process in terms of Fourier transformation is that if you Fourier transform twice you get back to what you started with.

This sort of 'sharing out and bringing back' aspect of brain processing seems very reasonable, whether or not it follows Fourier mathematics or simply has a similar dynamics. (There is an account of a conversation between Pribram and Fergus Campbell on a mountain in which there seems to be an agreement that the latter may be the case.) It provides some sort of order or hierarchy to the colony of neurones. However, on the basis of awareness in single cells it might seem to suggest that some cells would experience patterns as they are and some would experience them 'inside out' as holographic transforms. Although this might seem to create a problem I do not actually think it does. As I shall deal with in the next chapter there is no need for all cells to share a common language for patterns, all that is necessary is that the language is consistent for each cell. It does mean that different types of cell would have a different 'perspective' on the inside story, and I think that must make sense in terms of usefulness.

A more recent analysis of the mathematics of cellular interaction comes from Erhardt Bieberich (2002), who has suggested that fractal algebra may apply. The essence of this is in some ways similar to the holographic concept in that each cell is seen as receiving a pattern that mirrors the information in the net as a whole. It provides an interesting link between the more usual approaches to binding of sensations and that followed by Steven Sevush (2006) and myself. It suggests that a 'copy in each cell' of the global function of the brain is not only feasible, but is rather to be expected. As yet I am not aware of any experimental evidence to support this way of looking at things but it is nevertheless interesting.

The general concept of information about a pattern as a whole being distributed widely through the brain is attractive. It would fit with the idea of many cellular listeners receiving a single story, but at least in the traditional holographic view each cell will receive one and only one *component* of the incoming pattern. The component is a relationship rather than a pixel but it is still only a fragment of the story. Somewhere, there seems to be a need for a central 'hub' of cells receiving the story in the round. I shall come to this shortly.

A bunch of Bingo ladies

In addition to the idea of cells' access to information in the brain being ordered in terms of 'sharing out and bringing back', and perhaps 'turning inside out and back again', there is reason to believe that some information becomes 'privileged' in the sense of being the focus of attention. One of the big problems with a view of awareness as something global to the brain is that there seems to be no particular reason why some information in our brain should be 'conscious' and some 'unconscious'. It is as if some of the words in the book of our brain are *in italics*. Nobody has provided a plausible reason for this (except perhaps Steven Goldberg (2006), who also envisages multiple loci of awareness). If each cell is aware separately the problem disappears. Each awareness is limited to that information coming in to that particular cell. However, we do still need an explanation for the fact that a limited body of information seems to be available to those cells which control our speech, our considered actions and all those things that we associated with 'conscious behaviour'; what Goldberg (2006) neatly calls the content of 'customary consciousness'.

Some sort of mechanism for selecting things from different senses, memories and sources of emotion for a field of attention is needed whether awareness is ascribed to some global mindspace or to individual cells. As I sit I receive a constant input from many senses but I may select the sound of the fire crackling or the pressure of the arm of the chair on my elbow or the choice of the next word to be at the centre of my awareness. A circus needs some sort of ringmaster to guide the audience's attention to the next trick.

Some of the basis for this selection is reasonably well understood. When signals come in to the brain from the sense organs some of the cells they first reach seem to deal with sorting operations that do not seem to figure at all in our awareness. In the occipital cortex visual signals come in to an area called V1. They then pass on to an area called V4. It seems that only from V4 onwards is the information potentially part of the 'neuronal talking shop' that weaves the inside story. Signals passing along certain pathways in the brain seem by definition to be outside the awareness that we talk about. However, this does not address the issue of why the story is sometimes focused on vision and at other times on hearing.

A popular idea is that the signals that are in the field of attention are synchronised, often as regular pulses 40 times a second. This is unsurprising because if you want the information to be integrated in an ordered way you would want the timing of arrival of signals at cells to be precisely matched. What I find difficult to follow is the idea that being synchronised might somehow make the signals conscious. Admittedly synchronised signals might effectively lead on to further signals and non-synchronised signals might cancel out and come to a stop. However, I would then expect the events arising from the integration of synchronised signals to be the field of attention, not the synchronised signals themselves.

Whether or not synchronisation is important, and it probably is, it seems that we need some sort of mechanism that allows signals arising in different parts of the brain to 'compete' for the field of attention, and for the winning signals to be diverted into a set of pathways that feeds 'full story' awareness. How this might come about depends on where we place awareness. I tend to think of it in terms of a Bingo hall where the Bingo follows slightly unusual rules.

For non-British readers Bingo is the British version of Lotto or Housey-Housey, traditionally played by ladies of indeterminate age and unlimited enthusiasm. This analogy is not intended to be rigorous (it has fairly obvious flaws in it) but merely to give a flavour of what might be going on. Each player (a cell) has a game card carrying a unique selection of twenty numbers all between one and a thousand. At the centre of the hall is a curved screen on which numbers are displayed in such a way that different players get a different view of what is on display. One player starts calling out the numbers she can see. All players tick off numbers on their cards when they either see them on the screen or hear them called. As soon as a player has heard all her numbers she calls Bingo and takes over calling out numbers, on the basis of what she can see. The flow of numbers *being called* represents the field of attention.

The basic idea is that a limited field of attention is constantly shifting in response to useful or interesting patterns in the material just presented. The gatekeepers to the field of attention respond on the basis of a mixture of what is in the field and what 'raw data' are coming in from the senses at the same time. The role of gatekeeper shifts round, but in a meaningful way. Everything depends on timing; on who shouts first.

Perhaps the first query that such a model raises is what is receiving the information specifically in the field of attention. There is a sense that information needs to be passed 'on and up' to another hierarchy of cells not concerned with the gate-keeping process. This brings us back to the issue of how many cells receive the story in the round, the story as we talk of it as being in our awareness, and whereabouts in the brain they are to be found.

How many cells listen to the full story?

I have to admit that up to this point I have made it rather unclear exactly how many brain cells I am suggesting have access to the 'full' inside story. I have left the issue until now because I have two reasons for being cautious about trying to provide an answer. Firstly, I am not convinced that there is necessarily one right answer to the question. Secondly, even if there is, there is a limit to our ability to answer it at present which is just the same limit that neuroscientists have in being able to say which cells in which lobes of the brain receive information from which other cells in other lobes and how that information contributes to the way we behave. A lot is known but a lot remains to be found out.

Asking which cells have the full story of self presumes that there is one such story. I am quite clear that there is no more than one story *of a given type*, in the sense that we do not have one field of attention thinking it is Tuesday and another thinking it is Monday, although on waking the ideas of Monday and Tuesday might be jostling for acceptance within the single story. However, it is almost certainly true that cells in different parts of the brain deal with different aspects of the story. The brain is unlikely to be much like the six blind men who disagree about the nature of an elephant because they have only found different parts of it to examine by touch. It may be like a blind man, a deaf man, a man without sense of smell and a man without sense of touch and the elephant. Each will have a different description but there will be no two descriptions of the same kind which conflict.

If the mind is a luxury cruise liner travelling from Southampton to Bombay there will be one story of the trip but it will be very different seen from the eyes of the kitchen staff, the boiler room engineers, the captain and the passengers. Perhaps the full story belongs to the captain. However, if we are thinking of this cruise as a cultural trip dealing with ancient history, we may consider the captain's story no more relevant than the galley staff's story. We would consider ourselves to be the passengers drinking in the pearls of wisdom of our guides.

The point at issue is that the behaviour of a human being, including the discussion of awareness, would have to be seen as *dependent on all these versions* of the inside story. There may be no precise cut off between the 'sort of awareness we talk about' and the sort of awareness we do not talk about. Even discussing metaphysics much of our communication arises from information not available to 'the awareness that we talk about'. Nuances of meaning may be conveyed by inflexions of the voice or hand movements and only afterwards may we be aware what it was about our thoughts that they conveyed. As I shall discuss at the end of this chapter, the processes involved in talking about our awareness may be much more cryptic than we would like to think.

This means that it is dangerous to assume that interaction between human beings involves only the versions of their stories that belong to some privileged group of supervisory cells. Another illustration might be a conductor and choir, where each chorister is holding a full score of all the parts and using it to ensure their timing and dynamics are correct but the conductor does not bother with a score but uses his knowledge of the overall shape of the music to draw out a certain interpretation. The choir could sing quite well without the conductor. Are we interested in the choristers' story or the conductor's story? I think we need to accept that we may be interested in all these sub-stories in different contexts.

But coming back to the first point about answering this question, I think that there are reasons for thinking that there is a particular type of sub-story that we are particularly interested in which approximates to the story we consider to be in our field of attention. As I sit on this sofa it is quite clear that my brain, or at least my spinal cord, is using information, even if it is negative or neutral information, from the touch receptors throughout my body to carry out a range of boring tasks. In the last twenty minutes I must have moved my feet several times to ensure they do not swell up under the pressure of blood in the veins (you may not know you do this all the time in response to what would be a slight discomfort if you noticed it). I have scratched my ear no doubt. However, if touch is in my field of attention at all it is the touch on my fingertips of keys on the laptop which I have not quite hit surely and which may have resulted in a red sqiggly (sure enough) line coming up under a word.

You do not have to be a neuroscientist to think it likely that quite a significant chunk of the brain may receive information only relevant to the field of attention. We may have reservations about this; we sometimes realise that something out of our field of attention has brought back a memory that comes into our field of attention. Only on puzzling it out can we see what it was that triggered the memory. Even the story of our field of attention is slippery and flexible in extent. However, it is not unreasonable to think that the brain deals with it in a rather special way.

So what does neuroscience tell us about where the cells might be that receive the information that is our field of attention? A lot of interest has focused recently on the prefrontal region; a little bit back from the forehead and slightly down towards the temple from the midline. It has even been suggested that this area forms a sort of 'homunculus' in the Cartesian sense. Steven Sevush (2006), who like me is keen on the idea that awareness belongs in individual cells, has reviewed in detail the inputs and outputs of the left prefrontal region which is where he would put 'verbally reportable consciousness'. He concludes that all the sensations we are aware of, or report we are aware of correspond to inputs from appropriate areas elsewhere in the brain. Outputs to speech centres, and directly or indirectly, to other motor centres are also present.

Like all areas of cortex the prefrontal region has connections to and from the thalamus deeper in. It also appears to have a constant exchange with the parietal cortex a bit further back. The parietal region seems to be particularly involved in knowing how things fit together in the world. When the parietal lobe in damaged you may be able to feel and describe the hard edges of a coin or bottle opener but may be unable to recognise how they make up a recognisable object. The idea has been suggested that signals exchanged between the prefrontal area and the parietal area may keep the prefrontal area updated about the way the world is changing around it in comparison to the way it thinks it is moving through the world. What is not clear in this context is whether 'consciousness' lies in the prefrontal homunculus alone or in its dialogue with the parietal lobe.

There must be at least a hundred million cells in the prefrontal region but there are different types of cell and it may be that the sub-story we are looking for belongs to only one type. Intriguingly, an unusual type of cell has been described in the cingulate gyrus, within this area, by Nimchinsky (1999). This type of cell has only been found in human brains and brains of closely related great apes. For those who would like human consciousness to be very special it would be nice to think that this type of cell is where it resides. However, I am cautious about such an easy solution.

While many neuroscientists seem to favour the 'front brain' approach, with something like a homunculus in the prefrontal region, a few would suggest that the most crucial area for awareness is somewhere quite different. Jaak Panksepp (2002) has pointed out that the brain stem has a reasonable claim to be the key place for awareness. Whereas some see consciousness as likely to be in the cortex because the cortex is very new in evolution, and these people see consciousness as a new phenomenon, Panksepp would argue that consciousness is either very old or has evolved from something very old and therefore is likely to be found in an ancient part of the brain. You take your pick. I tend to side with Panksepp, at least on this argument. Perhaps the favourite area in the brainstem for a homunculus is a column of cells surrounding a central fluid filled canal known as the cerebral aqueduct – the peri-aqueductal grey matter or PAG.

An attraction of the brain stem as a place where we might find cells with the 'full inside story' is that release of chemicals into the spaces surrounding cells during emotion is particularly well defined here. It is also close to a network of cells that appear to control sleeping and waking, known as the reticular activating system. Wilder Penfield found that the region of the PAG was essential for behaviour we associated with consciousness. At least the left or right thalamus also needed to be intact but very extensive damage to or removal of the cortex is compatible with 'wakeful' behaviour.

Putting cells with the 'full story' in the PAG does not detract from the idea that the prefrontal region may play a crucial role in building a story of self-awareness out of a continual dialogue with the dynamic context provided by the parietal lobe. It might be that although the prefrontal region is not the place we are looking for it is the last but one place in the most complex chain of places feeding in to the real place. Put another way the versions of the story in the prefrontal region may be the most sophisticated 'inside out' holographic versions of what goes on in the PAG.

My reservation about suggesting that the PAG is the home of the full story is that I am not myself an expert in neuroanatomy. Steven Sevush, who is more expert in this field, is reasonably convinced that the full story must be in the cortex and probably in the prefrontal region. He has specific anatomical reasons for thinking this and I would tend to defer to his judgement here (Sevush, 2006).

We are in the realm of speculation but if we can meaningfully talk of a group of cells which have the sort of full story awareness that we talk about then I would guess that they are likely to occupy a chunk of brain containing somewhere around a hundredth of the total number of cells. Only a proportion of cells in a particular area may have this job so we may be down to a thousandth or even a ten thousandth of all neurones. That would be something like a million cells. This is a very rough guess but if pressed to give a figure I would go for something in this order.

A million cells? Should the global story cells be so multiple? If everything has been pre-processed in various other parts of the brain why not just have the story coming in to one cell? The answer I would suggest is that at every stage in the brain there is the same reason for sending a body of information to many cells at the same time. It means that patterns in the information can be identified quickly and with fine specificity. The point is that whatever cell we may think has a rich input that could represent our experience it must have a job to do. There is no point in setting up a rich pattern in a cell that does nothing. The way the brain is organised it seems likely that there is no job that is dealt with by only one cell. At every level, banks of cells are used to do a job in parallel. This is where my view of things converges with that of Dennett (1991). One thing we are agreed on is that there is no sense in having a single place where everything comes together. In Dennett's words, at every stage there are multiple drafts of the story. The difference is that Dennett seems to imply that these remain fragmentary, which would seem to be no use for a brain that wants to respond to a complete picture. For me the drafts in at least one part of the brain, perhaps the PAG, are multiple, but they are complete.

For me, this scenario of a hierarchy of different parts of the brain each with many cells doing a job in parallel is helpful in trying to sort out how our awareness relates to that of other animals. We can allow humans to be unique if we see the human brain as built as for simpler systems but with one or more further layers of hierarchy added on. If people want consciousness to mean thinking about thinking then it may be that you have to have specific prefrontal-type circuits to get things to work. Awareness in other animals may be real consciousness in a lesser sense, but there may be a further level for us. The mapping of the outside world and the language used in that mapping may have undergone a crucial transformation, perhaps the ability to map, or at least label, the process of mapping. Some peculiar trick in the circuitry of the brain seems to have occurred in the evolution of *Homo sapiens* that allows the system to refer back on itself one more time. The medium used by this trick is clearly language, although sign language shows that it does not have to have anything to do with sounds. Language has allowed us to share our internal experiences, the experiences of the insides of our heads, with others. It may also perhaps allow us in future to understand the whole mapping process and therefore to understand the innermost mysteries of the universe. I will explore this in a further chapter.

Theory of mind

When talking to Chris Frith about consciousness, he made the point that consciousness is something that needs to be seen in a social context. Coming at the subject from a different angle I found this initially a little difficult to know what to make of. Having read what he and his wife Uta have written I now see why this is so important (Frith and Frith, 1999).

Thinking allows the creation of an awareness that is not just neutral acceptance of information from outside. What most people call consciousness is a form of awareness that implies that the information comes in pre-interpreted in terms of a concept of outside and inside. The information is a map of an outer environment encoded in some internal signals. The information is also interpreted in the context of a sense of sequence, and related to previous experience. In the adult it is most crucially associated with the context of a concept of 'me', a sense of self; self-consciousness.

The sense of me has been shown to be closely associated with a sense of others, of other agents, like or unlike oneself, of other centres of consciousness. This is often referred to as a 'theory of mind', a belief that other people's heads contain minds which work like out own, feeding on information from the environment but from one point of view, isolated from other points of view.

The importance of this theory of mind, or story of self, to thinking is that it is likely that we have to think of our inside story not just as a selected field of attention, but also as a pre-interpreted story of a mind, and a mind amongst minds. The potential worry is that if our story comes pre-interpreted, how much, if any, of it can we trust? Is the idea that there is something in my head that has a rich experience merely a pre-interpreted illusion? My thinking is that we should not worry too much about this. I suspect that our built in theory of mind works well because it is consistent with the nature of the system it inhabits even if it is illusory to the extent that it tends to make us presume the story has only one listener.

And for practical purposes it would be entirely counterproductive for a cell to have a sense of its own identity. It could not really have done so until somebody, such as William James, raised the idea, based on metaphysical argument, that it might have to have an identity, but even now it cannot have such an individual sense of cell-self. The information coming in to a neurone is not about itself, it is about the colony of cells as a whole that is a brain, or indeed a human being. My proposed listeners can never have their own stories; that was never what they were there for and it would do them no good. They are there to listen to the story of a whole animal.

In fact, it seems likely that the idea of the whole body as a single self amongst other similar whole body selves is very deeply ingrained into the pattern of connections in our brains and may go back quite a long way in evolution. Monkeys have cells called mirror neurones that recognise movements of other animals as corresponding to similar movements in themselves. Thus the monkey has a built in mechanism for telling it 'he is moving the bit of his body which on my body I think of as my arm'. It is not clear that monkeys have a theory of mind in the sophisticated sense of being able to think 'he does not know where the food is, but I do'. However, it seems that the illusion that we are single living units extends to instinctive recognition of other copies of such a unit; a body.

What is clear is that our understanding of ourselves as single human beings is not something we have come to either by pure observation or by logic. It is an interpretation that is normally contrived by our genes. It runs in to trouble when we try and think of our internal workings. We have no conception of our brains until we see pictures in a book. The more we discover about the detailed structure of our brains the more we are puzzled at how it holds a sense of person. We know that something inside is receiving a concocted story. Consciousness is awareness in the context of such a sophisticated concoction. Consciousness depends on the complicated brain structures that do the concocting but this concocting is not part of conscious experience itself. We have no information available about what it is that listens to the concocted story. Historically it has been assumed to be the brain, but the brain must also be the concocting device. I am suggesting that we simply need to start afresh with thinking where the story may be heard.

Thus, the story that I am telling here is in no way in competition or conflict with Frith's viewpoint. He and others have made significant progress in understanding where the story is being concocted. I am simply suggesting a practical solution to the otherwise fraught question of where the story is being heard. The idea that there may be millions of concurrent copies of the story is no problem.

eleven

Talking About Awareness

Having considered various aspects of the link between awareness and thinking I need to consider the issue of how we can
talk about our awareness, particularly if, as suggested, there
are many copies of it. If I now write about seeing blue sky having just looked up for an example of something to be aware of,
which turned out to be blue sky, in what way does the experiencing cause the writing? It feels as if the experience caused
the writing, yet if each cell has its own awareness of blue,
including the ones that are tuned to red or grandmother, from
the viewpoint of most of my brain cells, it was *not their blue* that
caused the writing.

This does not stop the awareness of blue contributing to output in those cells tuned to respond to blue, but it makes the role
of awareness more subtle. It also makes it clear that it may be
very easy to build false arguments around the causal role of
awareness if it is not appreciated that there may be more than
one of them.

It might be thought that a cell responding to blue ought to be
aware of something different from a cell not responding to
blue. However, that is probably not what we should expect.
There is no advantage in a cell knowing what its output is. The
cell does not want to know about its Herculean efforts to integrate a thousand H-H waves and come up with the conclusion
'not my brother's nose' when what all the cells want to hear is
the consensus view; 'grandmother'. When a cell fires, its output almost certainly has to go through sifting, synchronising
and consensus-forming by other cells before it can feed back
into an element of the field of attention.

Even if we accept that our intuitive sense that the blue here is
causing me to write about blue cannot be firmly grounded, it

does seem to make sense that experience affects actions. Whether or not we accept this depends on just how we see the relationship between awareness and access to information. My view is that they are the same thing, because physical interactions cannot occur without 'experiences' of some sort by something. However, some people may be sceptical about this and say that whether the access to information affects output, which it clearly does, and whether the 'feel' of that access affects output are different questions. It may seem that it is the painfulness of a pain, not just the availability of a pain that makes me cry out. I tend to see this as making things too complicated. Nevertheless, in one sense I can see a way in which this might have an important physical meaning.

My model of what may happen in the dendrites of a neurone allows for output to occur in two different ways. It allows for PSP H-H waves *either* to add up electrically without any global listening wave contributing, or to be modulated by a global wave, with the possibility that, for instance, the one occurs during sleep and the other during waking. In this sense access to incoming information meaning blue may only be associated with an experience of blue some of the time. Nevertheless, I prefer to put this to one side and focus on the situation where we are awake and aware of blue and in that situation I tend to think that access and awareness are probably inseparable concepts.

I agree that there is a nagging sense that when we talk about experience our speech is affected by the 'feel' of experience, not just access to signals. It is difficult to construct arguments about it being the blueness of blue that makes as talk about blueness because it could be explained by access to a signal meaning blue. What may be more interesting is being able to talk about the richness of an experienced pattern. It seems that the fact that a set of signals are all accessible to the same thing can affect output. Yet in a mechanical device like a computer it would not matter whether two, four or ten signals are accessible to each processing unit, it would seem that you could write a program with the same input-output relationships.

If the H-H waves do not join in a single, bundled together, totally interdependent interaction could we use them to report the bundled-togetherness of awareness? Somehow the inter-acting H-H waves need to be able to spread around the mes-sage that everything going on in a cell works as 'all in one frame'. One possible explanation might be that everything happening in a cell takes on the meaning of 'all in one frame' because the brain works by sending out 'questions' to cells, in patterns of signals, to which the cell gives 'answers'. Perhaps the way language is constructed makes us consider everything going on in a cell to be 'all in one frame'.

But I find this explanation falls short because it allows the processes inside the cell to be whatever you like. It is compati-ble with the idea that what goes on inside a cell is a series of disjointed interactions or computations working very much like a series of gates in a computer. If things were like that it is difficult to see why any aware wave in the cell should experi-ence everything as in a 'single frame'. That may be the way the brain's language works but it does not seem to give a complete account of why we have awareness like we do. It should not allow me to *write down* the fact that over and above the issue of the brain handling what comes in to a cell as a package, I *know from experience* that my experiences really are all of a piece. I think there is a rabbit loose here. However, my intuition is that the puzzle reflects further false assumptions about the way language works and that when we can see through these the puzzle may melt away.

Nevertheless, for the time being we are left with the idea that our brains integrate very rich patterns of signals in cells in a seamless interdependent way not because there is an advan-tage in feeling seamless but because seamless integration has some operational advantage. Darwinian survival pressure should in some way favour having a brain that happens to integrate rich patterns seamlessly. One advantage might be speed. If the brain used serial computation in the way a com-puter does, recognising grandmother might take a week. But I suspect there is a much more important advantage. A system that processes elements of information all together may be *better at evolving to do it even better*.

The way the body produces and uses antibodies seems to provide a good analogy. An antibody is a molecule that recognises another molecule by interacting with several places on the target molecule all at once. All the elements of shape information on the target molecule are 'processed' in a totally interdependent way. The immune system evolves on a daily basis to produce useful antibodies by randomly changing individual parts of the shape of each antibody, testing out billions of different antibodies to see which one fits best and then manufacturing in bulk only the ones that fit well with any microbe molecules around. It then goes through the random changing process again and again until only the antibodies that fit perfectly are made.

A mechanism like this is apparently extraordinarily wasteful but gets results very quickly. Perhaps more importantly it may evolve towards perfect recognition more effectively than a more 'systematic' mechanism, if indeed such a more systematic mechanism could arise in a living organism. There would be no value in changing components of the antibody systematically if what mattered was that all the components formed exactly the right pattern as a whole. It would be a bit like trying to pick a lock by trying to file the tip of the key to the right shape first and then the next bit along and so on. The trouble is you cannot tell if you have got the tip right unless you have got everything else right. Antibody design does go through a sort of 'rough cut' and then a 'fine tuning' stage but at both stages the changes made are random.

It may be a toss up as to whether a grossly redundant random mechanism or a systematic mechanism might achieve a biological objective but there may be an even more fundamental reason why biological organisms make use of random mechanisms based on the chance finding of an 'all-at-once' fit. It may be that such a mechanism is more likely to evolve. Of interest, it is how genes themselves work. The evolution of a new organ or a new species adapted for a new feeding behaviour occurs by random change and the chance occurrence of a change that fits all the countless environmental circumstances that apply to the organism at that time. 'Give it a whirl and see if everything fits beautifully all at once' is the basic mechanism of life and evolution.

What this may mean in practice for the brain is that it uses a mechanism for information processing which is based on many elements being recognised in a totally interdependent way because that is a mechanism that can evolve easily and rapidly. If the brain dealt with information in a more systematic way, like a computer, it might be much less able to take advantage of the one in a billion chance that a tiny random change either in the genes or in the brain itself might just happened to allow the animal to do something new, like fly or talk. It may also be an important protection against competing biological interests. Biological systems often seem to include a capacity for random change because that enables them to outwit predators such as viruses that can take advantage of uniformity or systematicity in the host. *Homo sapiens* may have out-competed relatives such as Neanderthal man because superior random pattern matching in neurones may have thrown up 'inspired' ways of getting the food.

I cannot be sure that this explanation is as strong as it might seem. However, within the genes there is evidence that the system is designed not just so that it can evolve, but also so that it evolves efficiently. We are pre-designed to evolve. For instance, there is a mechanism for turning one gene into two. A lot of our genes are duplicates of another gene that have undergone random change until they do something new, different and useful. The bottom line is the idea that our brains are likely to process information as all-in-one patterns because that is a mechanism that is likely to have evolved because it is good at evolving.

This leaves us with the rather difficult idea that it might be possible to have an animal or machine that behaved exactly like us, including writing discussions about the paradoxes of awareness, but which does not integrate information in its neurones in a totally interdependent way and which therefore has no awareness of the sort we think we have. Curiously, in one sense that is what human animals are like in my view because the brain as a whole is precisely a 'zombie' of this sort, supported by millions of aware components. However, I think there may be a reason why no such animal or machine, with no awarenesses of our sort at all, would end up discussing its awarenesses. I think it may be that our ability to talk as if aware arises from a process whereby connection patterns evolve during childhood in response to rules of language that depend on constant interaction between the discrete message sending between cells and the interdependent integration of signals within cells. As I have indicated elsewhere I suspect that this process cannot be algorithmic simply because of its sheer real-time complexity. This may be why an algorithmic Turing machine-like imitation of thinking will never yield something that of its own accord develops the tendency to talk about the puzzles of its non-existent awareness. I am very aware that our current notions of how information is processed fail to provide a clear theoretical model to base this idea on. However, it is generally agreed that no mathematical model will handle semantics at present.

So perhaps it is right to say that our awareness is rich because it goes along with a way of processing information based on rich interdependence of signals. The rich *feel* of the awareness that we call consciousness has no biological advantage. It is just that awareness is everywhere and 'rich' processing has the advantage that it is good at evolving to be very clever.

In my two-wave hypothesis I suggested that a global resonant wave in a dendritic tree might modulate the output of action potentials from the cell in real time. I see that as the most likely possibility. However, there may be reasons for thinking that a global wave might have a more indirect and delayed effect. Several people have suggested that our awareness is not there to cause our actions *at that point in time*. Rather, awareness may be involved in the laying down of memory and making us ready to act more effectively next time round. In many ways it is our intuition that our awareness relates to those actions that we want to improve on, rather than routine actions we have got used to doing as well as we need.

The arrival of a pattern of inputs at a nerve cell can have two outcomes. The first is the firing of the cell, the formation of an action potential that passes down the axon and signals to other cells. The second is a re-adjustment of the sensitivity of the cell to future inputs. This process, known as Hebbian reinforcement, after Donald Hebb (1949), is not well understood, but it seems likely to involve changes in electrical potential and calcium ion levels feeding back on to synapses. There seem to be two stages of the resulting changes, one transient and one more or less permanent. The longer-term changes involve the number and efficiency of ion channels used by the synapse to produce an electrical wave when a signal arrives from another cell. Molecular structures known as NMDA receptors are specifically involved (Perez-Otano and Ehlers, 2005). Reinforcement might take the form of just having a bigger H-H wave coming out from a synapse. However, it might also mean a change in the local membrane structure that made it more likely that a future H-H wave would interact positively with another H-H wave from a particular synapse. This might mean changes in ion channels that did not simply increase the amplitude of the H-H wave but changed its time course in subtle ways, possibly even by causing the synaptic spine to change its shape or position. The shorter-term effect, which carries the 'memory' of the re-adjustment process until more permanent changes have had time to occur, is less clearly understood.

Something that has intrigued me for a long time is the potential link between the prion proteins that when abnormal cause Creutzfeld-Jacob disease and awareness and memory. Prion diseases in general impair awareness and memory. A rare form called familial fatal insomnia produces an inability to switch awareness off, an inability to sleep. Prion proteins are present in association with neuronal membranes. What is fascinating is that they are also found on the surface of cells in the immune system called follicular dendritic cells that are important for memory of foreign antigens in our immune responses. I was very amused and gratified to see that a 2000 Nobel Prize went to Eric Kandel (2001) and colleagues for showing that long term memory is in part mediated by a protein that behaves like a prion. I would like to think that maybe one day prions will link up with the idea of awareness in a cell, but for the time being that is just a wild speculation.

Hebbian reinforcement is a response to a combination of events. A synapse will be reinforced if it has received a signal *and* the cell has fired. This makes the cell fire more readily when that synapse gets a signal again. However, if cells are designed to respond to patterns of signals it would seem necessary for reinforcement to take into account not only the signal at the relevant synapse and the firing of the cell, but also which other synapses received signals at the same time, or in a specific time relation to the first synapse. This would seem to require something which co-ordinates the reinforcement process at many separate synapses. In effect we are looking for something that 're-tunes' the cell so that it responds to a pattern of combined inputs. This is where a role for a 'chiming' resonant wave in the cell membrane might seem to have an obvious potential role.

This links awareness directly to the gradual development of a pattern of behaviour of a brain that constitutes a 'personality'. It emphasises the role of awareness in memory and learning, in determining what happens 'next time around'. This fits rather well with certain ideas Valerie Gray Hardcastle (1995) and others have suggested – that awareness is bound in with a particular type of memory that requires recognition of detailed patterns in time and space, including specific episodes and events.

What I have certainly learnt in trying to tame this problem is that causation is a much more subtle thing then we tend to think. John Searle has said some very sensible things about this. Things happen for reasons other than one billiard ball hitting another. Bohm and Hiley's (1995) concept of active information makes it clear that causation does not require work to be done. A multiplicity of awarenesses further alters the arguments in a variety of ways. Debates about how 'mind' can affect 'body and vice versa begin to look simplistic. We may not have solved the problem of how we can talk about awareness, but we need to know exactly what the problem is before we can know how to find the solution.

twelve

Languages of Thought

The suggestion that our experience of the world should be just an inside view of local patterns of electrical or elastic forces in a cell membrane may seem ridiculous. However, I do not see that it requires any apology or special pleading since any other suggestion for the basis of our awareness would seem just as ridiculous. It is no more ridiculous than the idea that experience appears like a genie from a lamp inside our heads as a result of these same changes going on separately in lots of cells. It just brings us face to face with the problem.

And the problem this chapter has to address might seem to take us firmly beyond anything that might at present be testable by experiment. Nevertheless, people like Noam Chomsky (2000) have shown that interesting things can be said about languages without knowing the physical reasons. The adage that it is more important that a hypothesis be interesting than it be correct may be more interesting than it is correct, but it may still lead us down useful avenues.

How can the contents of our awareness be 'displayed' in a cell? Certainly, there is nothing like a flat screen with a little man 'looking' at it. It does not take much to realise that. The whole idea of flat images is an artefact of our recent familiarity with pictures and television. The human brain evolved to use a three dimensional sensory apparatus to convert a three-dimensional world into a three dimensional experience. You might think that each eye gets a 'flat picture' and then adds the 3-D after comparing it with the other eye. I think it much more likely that all three dimensions are analysed in parallel right from the retina. According to my view there may be cells that deal with and are aware of just up and down and side-to-side dimensions but my guess would be that flatness would not enter into either their structure or their experience since flatness implies the presence of a scale of depth that happens to read zero.

It is true that flat images like squares may produce activity in neurones laid out roughly in a square in the visual cortex but this is of no more interest to awareness than the fact that neighbours telephone lines go close together in a cable.

Second transduction, translation of electrical impulses into something else where they arrive, does not mean creating a copy. What I think it has to mean, however, is creating some sort of coded physical pattern that corresponds to the outside world in a useful way. The problem is that the code could be anything, as unlike the world as a spoken name is unlike a brass doorplate.

We must remember Russell's argument; all we ever see is the inside of our head. The room around me cannot be thought *to have the shape of a room* in any absolute sense. It has features which match up very well to the image in my head but that is only a fabrication, wherever it might be arising. Experienced space and physicists' space are two quite different things that have quite a good correspondence in certain respects, but no more than that. We should not be puzzled by the fact that physicists' space is curved by gravity even if we cannot tell that it is. This is an aspect of physicists' space that experienced space is not built to reflect.

Moreover, to avoid the infinite 'Russian doll' regress of needing a viewer inside a viewer inside a viewer and on for ever, this fabricated image must be part of something that views its own domain, in essence that views itself. It might be argued that the view we have of the room is not what the inside of a cell ought to look like. But what ought it to look like? Whatever this map of the real world in our head, (that we take as the real world) is made of, we have no reason to suppose that we should have any idea what it 'ought to' seem like in an awareness.

I seem to be faced with the stark question of what variation in a membrane might be the taste of cheddar cheese and why. At first this seems an impossible problem, and it may be that to an extent it is and will always be. However, there is an approach developed by philosophers Kevin O'Reagan and Alva Nöe (2002) that may help take us part of the way to an answer. It is suggested that the elements of our sensations, the qualia, like red and cheddary, seem the way they are because of the way they relate to other things and in particular to what we are doing at the time. For instance, visual sensations are those that change in a certain way when we move our head. Cheddarness involves messages coming from taste-buds, the nose and touch receptors on the mouth and tongue, together with information from inside our tongue about how we are moving it. In the absence of a sense of texture, which needs this knowledge of our own movements, a lot of tastes become unrecognisable.

The full approach, known as enactive consciousness, is a conjecture that leaves many questions unanswered, but it does seem to grasp the one handhold that we might have. It suggests the way to know what sensations are is to *know how they are interdependent,* how they interrelate with other inputs, including information about the brain's outputs. Arguably that is the only way we ever know what anything is and I shall make as much use of this handhold as I can. O' Reagan and Nöe suggest that awareness arises directly out of our sensory and motor interactions with the outside world, with no inner 'receiving unit(s)'. However, I think there does have to be a site of reception, which means the relationships that determine the type of experience must be reflected in relationships between inputs.

How many languages?

So, musing over the language or code in which qualia might be written, some general thoughts come to mind. It may be useful first to consider whether the language needs to be consistent between observing units, to be universal. Is red coded the same way for all cells in all animals? Are we looking for one language or some general rules?

The immediate answer to this would seem to be, maybe disappointingly, that there is no particular reason to think that the code for red would be the same in different cells, or even the same for a single cell over a period of time. There are two reasons for this. Firstly, if a cell is aware of red, that would be because a chemical signal or set of signals has activated the membrane at one or more synapses. The experience of red is then entirely a private matter for the cell, which may lead to the production of a further signal that is passed on back in the chemical code. Since awareness is never shared it does not have to match anything else. The cell's listening language is only for its own use. Cells might use the same language because it is likely to be roughly the same if they are built and connected up much the same way, but there is no necessity.

The second point is that a single cell does not even need to keep to one code for red because all that is required of the incoming story is that it is consistent at a given time. If I ask myself if the red in the picture is the same red as yesterday I sense that my memory of the red of yesterday is being fed in in the same code, so it has to be the same. We need to keep this in mind but I rather doubt that we need to worry about experiences drifting about through randomly varying language. What may be more relevant is that it may be naïve to think that there would be a code for red as such. We may always be aware of red as a quality of something, a relation to something else, a part of a composite, something which is always interdependent on other aspects of the flow of the inside story. We think of red as an elemental sensation but internally it may not be handled that way.

The implications for the commonly asked question 'do two people see red the same way' may be complex. The same electrical pattern may be used by most people in their mapping of red. However, this is likely only to be true for neurones at equivalent points in the brain. Neurones fed directly by red-sensitive cones in the retina may get much the same signal and have much the same experience as neurones fed directly from cells sensitive to middle C in the inner ear. The signal would only need to mean 'yes', because it does not need to be distinguished from another modality. The redness that we discuss should only belong to cells that have access to several types of sensation. This sort of redness is then in a language that crosses modalities. Thus, completely different mapping systems may be useful for different neurones. On the other hand it would be simpler, and perhaps make some functional sense, if most 'higher level analysis' neurones used a mapping system with some common ground rules.

It seems that even the ground rules for a language of awareness are uncertain, but at least we may be able to sense what this uncertainty might entail.

Bubbles, topologies and multiplication

There is a sense that qualia cannot be incomplete. Whereas my old laptop had a dud pixel that showed black amongst the sky blue desktop display just to the right of these words, the blueness of a clear sky is immaculate (literally, without spots). Our view of the world is like that, it has no grain, not because the array of receptors in the retina has no grain, but because qualia have no grain. Although we tend to think of visual images in terms of pixels this is a very new viewpoint. Even though we now have digitally encoded sound the code is at a level too fast to have any sensory equivalent, so the indivisibility of a sound is something we still tend to take for granted.

This sort of completeness makes me think of the analogy of a soap (thought?) bubble, something that only exists if complete. Perhaps at the level of a wave experiencing its domain, patterns exist which draw on the same mathematics as waves in a soap bubble. A bubble can jiggle in each of three dimensions in various ways, including turning into a dumbbell and back to a sphere. There might be equivalents of colour rings on the bubble. The analogy does not go far, but it may be a start.

I drafted the last paragraph in early 2003. This interests me now, in 2006, because at the time I was just beginning on this story and tended to think that awareness might be in microtubules, as Hameroff (1994) had suggested. I came to the idea of the soap bubble because my efforts to think what the mathematics of a thought might be led me to the idea of spherical shells moving out from a centre, complete from the outset but capable of complex variations. But I now see it as much more likely that awareness is based in the cell membrane, which is a smectic mesophase; a soap bubble. The coloured rings I had dreamed up are waves of thickness and thinness. The dumbbell jiggling is another wave. Even the child's playtime soap bubbles are full of waves. I would not claim that my initial thought of a soap bubble thought was more than a coincidence, but little comforts are welcome when trying to build an impossibly difficult idea.

In a computer, complex meanings that might imply several dimensions, or degrees of freedom, are shepherded into strings of signals with one degree of freedom. In speech two degrees of freedom, pitch and time relation, are available. As pointed out by Oliver Sacks (2000), the Sign Language of the deaf has all four dimensions at its disposal, and so is often very economical and expressive. This leads to the idea that sensations and perceptions may be built not by making strings of things but by making use of patterns in different dimensions, with the obvious choice being all four.

A wave pattern in a cell membrane should be able to make use of all three spatial dimensions and time, although as tube-like structures the dendrites of a neurone may support linear rather than drum-like wave patterns to a degree. However, from the point of view of awareness I am not sure how relevant the architecture is, in that for a field, taken as a single entity indivisible in time and space, all elements relate to all other elements, not just the ones next door.

The philosopher John Searle (1997) refers to the idea of consciousness being built up by modulations of a basal state rather than the addition of 'bricks to a wall'. His view of consciousness is of a field involving many neurones. I think I am proposing much the same sort of thing. I would simply suggest the shift of viewpoint to the idea that the field is the membrane of one cell, perhaps in interaction with the cytoskeletal framework.

Apart from the idea that observation in a field would not be achieved in bits, but as 'indivisible multidimensional topologies', there is the obvious possibility that awareness reflects change, salience, or differential. Redness may be 'more red than before' or 'more red than there'. Everything may be relative.

What worries me most about the idea of awareness in individual cells is the implausibility of the sort of richness we seem to experience even in a cell membrane (richness anywhere else seems to make no sense for reasons given). It is not that there are not enough inputs to a cell. What I find difficult is to see why a cell should bother with a rich experience if all it is doing is checking off whether the input deserves a 'yes' or a 'no' output, which, put crudely, is all we think that nerve cells do.

One idea that might seem to help is that cells with the 'full inside story' have the job of detecting shifts in relation to what the brain is expecting (as for O'Reagan and Nöe). Like bank note checkers at the Royal Mint they may be comparing two complex patterns, perhaps coming in in parallel or rapid sequence and picking up discrepancies. Maybe you need the present experience laid out in all its complex relationships so that you can make a comparison.

A language for space?

The more I think of it the more it seems to me that for cultural reasons we may have wrongly come to assume that any representation of the world there is in our heads works like a picture rather than like words. We are used to the idea that words are just tags rather than models of the outside world like pictures. Surely if the cell was going to be aware of something it would not use words to set up its representation? This may be so in some senses, but it is interesting to think what the most basic information might be about an environment that could be mapped to be available to a wave.

What about: here, there, right, left, before, after or even near to me, since we are assuming we are dealing with a self-aware subject? Pictorial attempts at these tend to be just as much tags as words, like arrow signs for instance. How many pixels is 'over there'? If a wave is having to 'make sense' of things it has access to, I am not sure there is any logic in thinking it would need the sort of input our eyes get. This is the basis of the homunculus fallacy, the idea that there is a little screen inside our heads.

And it intrigues me that I can describe the strange features of my image of creeping cinquefoil, with no number of fans of five leaves, in words, but I would be hard pressed to put it across in pictures, even though it is a picture.

Although I like to think of myself as a 'right hemisphere' visual person, mistrustful of those who glibly use words for things of which they seem to me to have no internal picture, and thus no understanding, I am tempted to think that, deep down, awareness may use a language much more like words than pictures. But those words can make pictures as well as more words. The words may also behave like the terms of a mathematical equation or a computer language, and I will come back to one of those. Maybe in a sense the physicists are right when they say that their differential equations *are* the reality, rather than the reality we naturally think of. That would suit me as long as the equations belong to packages of instructions that come in quanta. My reservation would be that equations might also be created and presented in ways that are not our reality, like in John Searle's Chinese Room.

Another major difficulty in getting to grips with the idea of how awareness relates to a physical structure like a cell membrane is that we tend to think we are matching up a subjective picture with a 'physical' structure of a cell based on what we have seen, or seen photographed, down a microscope, or in diagrams. The problem is that our concept of the structure of a cell is built in the same subjective language that we are trying to compare it with. Somehow we have to view the problem 'from the inside out' and accept that space may be mapped in a totally unexpected way, using whatever dimensions (if that is what they are) suit the purpose.

Just a hint of the way codes can transform things in unexpected ways comes from the way an x-ray beam passing through a three dimensional crystal produces a series of rings and dots on a screen that carry the three dimensional information in another form. Suggesting that the language of awareness is closer to words than shapes even for space may seem to go way beyond this, but I suspect that that only reflects the way we keep our thoughts in rigid compartments and fail to see their similarities. Transforming information in strange ways is now commonplace. Both sounds and pictures can be turned in to patterns on a reflective strip on a CD, patterns of magnetisation on a tape or frequency modulated (FM) radio waves passing through space. There seems no reason why cells should not make just as clever transforms. The key difference is they have no need to turn the transform back into a picture or a sound to enjoy it. That was the true homunculus fallacy.

A further mistake might be to separate space from its contents. We may not be aware of space so much as aware of things in spatial relation. These things may include sounds and smells as much as colours and shapes. When I hear a bird singing I do not necessarily separate the sounds from where they seem to be coming from. Even emotions may be in space in the sense of being at the centre of it. The more one goes in to it the more the idea of a simple spatial map for space seems unlikely.

Another issue is the illusion of detail. We think we are aware of everything around us, but I am virtually never aware of the weight of my shirt on my left shoulder, and only sometimes of the whistling tinnitus in my right ear that has been there for twenty years. I can be aware of them if I decide to be, or if my shirt catches on a door, but I suspect that I am actually aware of rather few things at a time. We think we are conscious of huge banks of information because we are used to the idea of images being made of huge numbers of dots or sounds of huge numbers of digital units. But it is not like that inside.

We are usually aware that what is around us is a familiar but very complex scene. I suspect that most of it is simply coded as 'usual mess' or 'rather streaky' or 'very smooth'. (That does not mean that 'usual mess' does not come with recording of thousands of bits of information in some other neurones and the availability of this information for downloading into the cell that has the experiences we talk about.) The number of things we NOTICE may be less than 100 per second; a mixture of some very simple raw data, some things which we have chosen to scrutinise and things which are at odds with memory. This is within the scope of a single cell. If lots of cells were involved in an awareness we would need to explain why we are not aware of the input of the other billions of synapses.

Why do we see no grain?

I have alluded to the absence of 'grain' in our visual awareness. Some years ago I was interested specifically in the absence of grain in our visual image of the world as part of my bachelors degree in Fine Arts. Why is it that we cannot see the individual dots in front of us even if we deliberately focus on the tiniest detail? For a start the comments given above might make one doubt this need be a problem. Dennett (1991) quotes an optical illusion in which you recognise the outlines of contrast between coloured shapes but 'see' the colours in the wrong places as evidence that we are not 'seeing' at all, i.e. there is no screen of pixels. But if awareness uses 'colour contrast here', 'even colour here', 'this colour red' 'this one green' as its elements rather than 'red pixel', 'red pixel', 'green pixel', the illusion is understandable. This would be how real seeing works. The mistake is to think that there is some other earlier dot-based seeing at the retina.

There is another explanation that is relevant to Seurat's painting. There are times when we do see dots, when the retina is only just coping with the information coming in. A good example is a very bright sunny day, when, walking amongst trees giving dappled shade, we are constantly moving from very bright to dark. Our visual fields are full of after images which some of the time obscure what is in front of us. Where they mix we may see showers of coloured dots, at least with a little introspection. Another situation is at night in a street with bright lights illuminating patches of colour, as in a fairground or carnival setting. Again our retinas are confused by light and dark and by the presence and absence of colour in bright and shaded areas. A third situation is by the sea, when mist makes half-seen shapes appear and disappear and even seem to move about as poor contrast and lack of reference points distort perception of space. These are the subjects of Seurat's masterpieces, La Grand Jatte and La Parade and a series of landscapes of the north French coast. Seurat explored our vision way beyond the optical mixing theories of his time.

So yes, there are times when our vision contains dots. However, these dots are evident only because they *contrast* with the colours next door. What we do not see is dots within an expanse of blue sky, and for good reason. To see a dot we must see a border round the dot to know it as a dot, in the way that we see a border around a dot on a computer screen if we get close enough. The question then arises as to why we do not see the border broken up in to dots with borders. There is an infinite regression. Seeing dots within a uniform field implies seeing more than the field itself. It implies seeing the boundaries between the elements in the field, which do not exist. We should not expect to see dots because we should not expect to see the boundaries between dots. There is a fundamental logical error in thinking that we should see dots at all. To think that is to misunderstand what sort of language of qualia we are looking for.

You might say that because each retinal cell has to sense light separately it should send an image to the brain in pixels. However, even within the retina much of the raw information is discarded leaving only messages about contrast or even the changing contrast of movement. If the information coming in to the brain is already in this form, what would be the point of converting it back into dots?

A language for time?

The idea of space being mapped is familiar, and, as I have suggested, perhaps misleadingly so. A picture of a big space in a small space on a sheet of paper is what we usually mean by a map. Mapping of time is a bit more difficult. I cannot think of a common situation where we map a long time into a short time except perhaps time-lapse film. We tend to map time into space, like in a concert programme or a diary. We even map time into space to read a clock.

Perhaps the fact that our brains map time into something else is best shown by our perception of movement. Even in the retina there are cells that pick up changes in patterns of light that generate a signal 'moving left'. Wherever in the brain we have the sensation of a black blob moving left this sensation is not based on a black blob here *followed* by a black blob further left. It is based on the two simultaneous messages black blob and moving left. Having a sensation of movement does not require any more time to elapse than a sense of red. Again, it seems the messages that we build sensations out of look as if they ought to be more like words than pixels on a screen.

There are other reasons for thinking that trying to find a language for time may require careful thought. Our perception of time is paradoxical. We seem only ever to be aware of one point in time, yet this point in time can, as pointed by E.R. Clay, famously quoted by William James (1890), seem to contain all the notes of a bar of a song. Clay called it the 'specious present'.

Physiological studies also indicate that 'now' does not relate to one point in time in our brains as a whole. If things happen in different places around us which provide signals for our sense organs which take different times to get to our awareness then some jiggery-pokery goes on inside to make them all seem to match up the way they should. You can catch the brain out by stimulating it directly with electrodes. The brain shifts the sensation by the amount it usually does to make a story that fits together. Now is an animated cartoon created by ten billion graphic artists, not a photograph or a view through the window.

The fact that now is created from things happening at different times in different parts of the brain is for me another inescapable argument against a global awareness made up of things happening in separate parts of the brain, bound by 'function'. There is no rational explanation for things happening at different times seeming to happen together. It highlights the problem with the idea that synchrony of nerve cell activity binds everything together. It is difficult to see how a 40 Hz cycle can bind together events that are out of step. Just as it seems essential that if you are going to act in response to a specific combination of sensations those sensations are going to have to meet up somewhere, it seems essential for sensations perceived as simultaneous to be re-presented somewhere at the same time. And presumably the whole point of the time-shifting mechanism is to get things synchronised when they do come together.

There is also a basic physical problem with the idea that sensations are bound by synchrony of classical physical events like nerve impulses in separate places. Special relativity tells us that 'now' is different for different places in space. Although relativity does not have much effect on anything we can measure in the brain there is a theoretical problem. However tiny the difference, the physical present of the right side of my brain is just not the present of the left side. Relativity does not allow an observer to have a present in two places at once.

This at first sight seems to suggest that the whole idea of a physical basis for awareness is impossible. One awareness at both ends of a cell at once might violate special relativity much less than opposite ends of a brain but it would still be a violation. However, my thought is that if we see our internal listener as a mode of oscillation or wave according to the rules of quantum theory this problem may go away. The package of rules that is a quantised mode has special relativity built in to it. Modes may take 'lots of different nows at once' in their stride. If a listening wave mode occupies and experiences a domain in space and time in such a way that there is no meaning to considering one part in isolation from the other parts then the problem seems to become a non-problem. Externally observable evidence of interaction between the mode and other things will always relate to local appearances and obey relativity.

Even I find the idea of single cells being aware pretty improbable when I try to imagine it, but it does seem to get round some very awkward impossibilities.

This way of looking at things makes me think all the more that it would be wrong to consider the language of awareness of space as being in space or that of time being in time. As one would expect from a wave, space ought to be mapped into relations of space and time and so should time. I have suggested that I think that what appears to be an instantaneous experience is likely to be the history, or sum over histories, of a wave occupying a finite, if small, space and a finite, if brief, period of time. However, I think it most unlikely that a time sequence in the outside world would be mapped into a time sequence within this history. To think that this might even be appropriate is, from my perspective, to slide towards the true homunculus fallacy of thinking that waves see like eyes or hear like ears.

In some respects our experience of 'here' seems much more extended than our experience of 'now', in the sense that the present seems to be instantaneous (with the proviso about notes in tunes given before). I am not sure that I have any good explanation for this. However, it may reflect the fact that information about the world tends to come in at the speed of light, whereas inside the brain things happen much more slowly. It may be easy to condense space but not so easy to condense time, and danger tends to require responses to be as quick as possible. Building up an awareness over a period of time could be very costly. It seems that we get round this by 'picturing history' in a different way from our sense of now. This may be true of awareness in all nervous systems but I slightly wonder whether the three-toed sloth has found a different solution. Its now seems to last for never less than ten seconds, even when surprised by an intruder!

Galen Strawson has suggested that people may fall roughly into two groups, those who see themselves existing only now, as one of a series of episodes, and those who see themselves as existing in a continuing narrative. The first he calls episodic, the second diachronic. What this may mean is that the human sense of extended time is quite an uncertain, flexible thing, perhaps because it has evolved under conflicting pressures.

As indicated in a previous chapter, I think it most likely that awareness comes in episodes of a fraction of a second within which there may be various references to now, the past, the future and the way things are progressing. I suspect that these episodes are shorter than the periods of a few seconds that Strawson considers to be allocated to particular 'thoughts'. The length of the episode would, of course, not be part of the story encoded in the episode, so we have no way of knowing how long these episodes last. Some people have difficulties with the idea that there awareness may come in very brief bursts but I do not think we can possibly rely on the intuitive

impression that we are continually aware. We have no awareness for the period when we move our eyes and our visual
field lurches to one side. Awareness is a much more artificial
thing than just a real time copy of the outside world.

Elision over time and the misconception of conscious will

It may be worth adding at this point that the way we map time,
the way that now seems to be an elision over a period that
includes a tune or a movement, may shed light on a common
confusion about what might be meant by 'conscious will'. As I
have discussed, conscious experiences are almost certainly
associated with events that lead to actions. However, there
appears to be an idea around that we can 'act consciously', that
we can experience our acting as we do it. There are good reasons for thinking that this is muddled thinking, as both Neil
Levy (2005) and Sean Spence (2006) have suggested. If we
assume that deciding to act involves an input leading to an
output then awareness, or consciousness, will be of the input.
The idea of a decision-making locus being aware of its output
directly makes no physical sense. Awareness of output
requires information passing round and re-entering the locus
as input later, whether the locus is a cell or a net of cells or
whatever. The sequence must be (1) awareness of the information on which a decision is to be made, (2) making of the decision (sending an output), (3) awareness of the existence of that
decision through new input. Consciousness of action must
always follow action. Thus, as Neil Levy points out, to be surprised by the experiments of Benjamin Libet (2002) is, in a
sense, to misunderstand neurobiology.

I suspect that this confusion arises precisely because of the
elision of events that we find in our sense of now. A single
frame of the cartoon story that the unit receives involves both
(1) and (3) and an account of their relationship. To be meaningful (3) must arrive with (1) embedded in it. Input is pre-interpreted as a sense of causal sequence. Thus, the elision that
allows one blackbird to hear another's song, that allows us to
use language, and that gives an account of sequences involved

in decisions may be fundamental to the mechanism that gives us the illusion of being a single living being.

Patterns as a basis for meaning

One view of our perception of reality is that it is less a question of perceiving what is out there now and more a matter of bringing up pre-formed 'hallucinations', or at least some sort of pattern of hallucinatory modules, to order. If awareness is in a cell we need to know how the cell is tuned up for such a pattern in advance. This relates to the spectre, at least as the fashionable cynicism would have it, of the 'grandmother cell', but also to the fact that there could be two quite different sorts of grandmother cell.

Like so many things in the study of the mind the term grandmother cell comes with various different assumptions. I have even seen it confused with the idea of an all seeing 'mind cell' more usually called a pontifical or queen bee cell. This is not what a grandmother cell was originally supposed to be. The grandmother cell is simply a cell that fires off a signal if, and only if, grandmother comes in to view. Whether or not such a cell is likely to exist is a matter of debate.

Horace Barlow said to me that he thought that it would be unlikely that we would put aside a cell specifically to recognise grandmother since there are a lot more important things to do. I would personally have thought that if we have a few billion cells to work with assigning one to granny is reasonable, but I concede that Barlow may have more sophisticated reasons for being sceptical. I am not aware that anybody has ever obtained experimental evidence for a grandmother cell but there is evidence from recordings made using electrodes in conscious brains for cells that respond specifically to certain objects, examples being pictures of furry rabbits and of Jennifer Aniston (Quiroga *et al.*, 2005). The concept is the same whether or not it is granny or a rabbit in question.

Having cells that recognise specific patterns seems fair enough but, as I have touched on in an earlier chapter, there are two ways this can come about. A cell recognising a pattern which involves the presence of elements A, C , G and J might just have one input for A, one for C, one for G and one for J and no others. The alternative is that the cell has many more inputs and recognises A as a combination of inputs a certain distance apart in space and time and C, G, J similarly. The second cell would have many inputs which at any one time were not doing anything but it would be able to recognise the essence of A, C, G and J in a variety of different contexts. It would recognise a pattern.

It seems to me that for the first sort of cell incoming information will only have one of two meanings: enough elements or not enough elements. Whether or not it is enough elements to recognise grandmother or enough to recognise a rabbit is of no interest *to the cell*. It is enough of the elements that cell deals with. For the second sort of cell information can have many meanings, including grandmother, rabbit, old lady, face, or furry animal and also jam jar, motorbike, or sunset. It would make sense for the second cell to have an awareness like ours but not for the first. Both cells would only respond to the subject they were tuned to but whereas it might make sense for the first cell to have tens or possibly hundreds of inputs, it makes much more sense for the second type of cell to have very large numbers of inputs to catch a pattern in as many ways as possible.

You might think that a cell tuned to motorbikes should not have the experience of recognising grandmother but this is where we need to remember that the experience of recognising grandmother is part of a re-circulating story. There is no reason why a functional unit in the brain that recognises grandmother, whether a cell or a network, should be informed that its activity, rather than the activity of some other part of the brain, was responsible for the recognising. No theory of the brain has a mechanism for something 'seeing itself deciding something'.

The point I am interested in is that wherever awareness is in our head it ought to be where many elements come together and where, at any one time, lots of other combinations of elements might have come together but do not. For a representation to mean something *at the place of representation* it must be not just the presence of a pattern but also the *local absence of many other possible patterns.* In a computer there does not seem to be any option for complex meaning at the place of representation because no more than two elements of information are present together in any one place. Brain cells are different.

The two types of cell I have suggested are not black or white alternatives. It is quite possible to have a cell that requires enough inputs of a certain type but which have to form the right sort of pattern with other inputs. Since the brain is at liberty to make use of whatever rules it can encode in its genes it seems likely that summation and relation are mixed in each cell in whatever way is useful at whatever level of processing hierarchy.

It does seem likely, however, that if the awareness we discuss relates to individual cells these are likely to be cells that are particularly pattern dependent. They may even be tuned purely to changes in patterns, to detect patterns of drift in experience – the most relational patterns of all.

Do we think in C?

About twenty years ago I came across the computer programming language 'C'. Unlike computer languages that come in lists of (1) Do this (2) Do that (3) Do this again etc. C is written, as I remember, using brackets. If I remember incorrectly it does not matter because it is the bracket principle that I am interested in. The idea is that whatever you have said the computer should do so far 's(g(d(i(k))))' then the next step will tend to take the form j(s(g(d(i(k))))) which means 'consider the relation of j to everything you have worked out so far'. It may be that s is in fact c(d(e)) so that you can have structures like j(c(d(e))(g(d(i(k))))). The brackets are a convenient way of telling you in which order you relate one thing to another.

Since in a computer you are always relating only two things there is only one term in each gap. However, if you are relating patterns you could have j.k.l.(c.d.e(g.h.i.(o.p.q))) meaning 'consider the relation of the pattern of g, h and i to the pattern of o,p and q, then consider the relation of the pattern c, d and e to the previous relation, then do it for the pattern j, k and l. There are all sorts of reasons for thinking that the brain might be doing something like this. The mathematics of such a language would be closer to matrix algebra than that used in a computer. Perhaps this might be a clue to the route to reconciling meaning with physics.

An interesting offshoot of this idea is that in some sense, for some cells, the first term might be considered to be 'me' or 'the self'. This means that the sort of code above is a continually repeated command 'consider how what has happened up until now relates to me', or 'listen to my story'. All previous versions of the first term will then contribute to each of the nest of following terms in a 'self-conscious' way. What was me a moment ago becomes part of what has happened until now. There will always be a self-term at any stage but it does not necessarily have to remain the same.

This way of looking at things may have some bearing on the comment made by Hume (1999): 'when I enter most intimately into what I call myself, I always stumble on some particular perception or other, of heat or cold, light or shade, love or hatred, pain or pleasure. I never can catch myself at any time without a perception and never can observe anything but the perception.' The implication is that the first 'me' term can never appear as the object of a relation, except in retrospect, or as an abstract construct. As the first, subject, term it can never be 'found'.

Where does Chomsky fit in?

Somewhere in the language of awareness we find a way for meaning to relate not just one thing to another but also things to language. Somewhere the concept of a physical cow has to be related to the word 'cow'. I have a feeling that this relation ought to have something to do with the relation of experiences to 'me', to have something to do with a more general relation that we might call belonging. Experience often seems to contain a combination of words or images with the respective things that belong. A familiar face comes *with* the individual features in their colours and tones, a tune comes with the individual sounds, a word comes with an image, sometimes only as a half-experienced shadow, sometimes in solid form.

I wonder if our ability to be self-aware and our ability to use language reflect a similar property of our nerve cells. Perhaps the way our cells relate patterns is just that much more layered than even in monkeys or cats so that, using the sort of 'bracketing' indicated, 'to me' and 'means' emerge. A basis for bracketing one sort of belonging in another in a single cell would seem to be within the range of possibility. Incoming signals may relate to each other in a hierarchical way.

This sort of discussion of hierarchies of meaning or belonging goes back a long way. Hume (1999) talked about the relation between an idea and an impression, the idea being the core concept (cow) and the impression being an example in current experience (this brown cow). One thing that puzzles me is that elsewhere Hume claimed we could only relate things in three ways. He claimed that things could only be related either by resemblance, by being close in time or space, or by one causing another. Meaning and belonging seem to fall outside this list. In a related but subtly different analysis Nicholas Humphrey (1992) has contrasted a perception (Frisian cow) with a sensation (one of various patches of black and white in a cow pattern). Somewhere in the brain these things must get related and 'getting related' seems to imply the togetherness that only occurs in a cell.

John Searle (1997) has emphasised the apparent incompatibility between semantics, or meaning, and physical processes such as we see in a computer. Perhaps we can say a bit more about the character of awareness than just richness. It also involves things like belonging. I just wonder whether this is somewhere where wave modes come in again. 'Belonging' has no obvious meaning in Newton's billiard ball physics, but it does seem to have a place in a wave-based universe. A quantum of momentum belongs either to this mode or that. A mode of vibration belongs to an aggregate of molecules in a crystal or a membrane. Belonging seems to be where two things share some element of identity while remaining separate entities, often of different kinds. Is it conceivable that this is where the semantics of our subjective world comes from? I only mention it as a passing thought that might one day grow into a hunch.

Noam Chomsky is famous for having put forward the idea that our brains are set up with the relations between words in language in place long before we learn our first word. We do not need to learn how a language works, how subjects have predicates and so forth, our brains already know that. Almost all we need to do is learn what the individual words are in English, Spanish or Urdu (and some conventions about word order) and we can construct sentences. There must come a time in the future when the physical basis of this pre-knowledge of language is made clear. It may only become clear if we understand the way things are related in individual cells.

The binding problem, which I have described as the problem of how we have access to a pattern in a 'single frame' is sometimes described in more specialised terms; how we link an element of information to one thing and not another. The question is how, when we are aware of a yellow square and a blue circle the yellow gets to go with square rather than the circle. Although the general problem of how anything goes with anything is my central interest this more limited problem is also fascinating.

What seems to me of particular interest is that in our awareness we have both raw data, Humphrey's sensations, and ideas, Humphrey's perceptions, to which the raw data belongs (Humphrey, 1992). We see a car but also the shapes and colours of the car. Moreover it seems that these things can be present together in a way that has no 'normal' meaning, as in dreams and trips on hallucinogenic drugs. Even in normal experience, as indicated in earlier comments on weeds and Seurat's paintings either sensations or perceptions can be null or misappropriated in various ways.

Trying to make sense of this cannot go much beyond guessing, but I do wonder whether the signals encoding sensations and perceptions are fed in parallel to cells in such a way that the cell can be aware of both and the different sorts of relationship they represent. Sensations might broadly be considered physical relations and perceptions functional relations. How these might be encoded is hard to know, but I come back to the idea that perceptions, ideas, and everything that we consider to be the higher-level mental elements that might make us different from animals, and which we tend to associate with language, are critically dependent on the elision of a sequence of elements into a single present. Mapping of a period of time into now is crucial to the human condition. Other animals must have something of this ability but perhaps language arises out of an ability to apply this process repeatedly; to create a now within a now within a now.

Is emotion a map of something?

A long-standing debate exists as to whether or not some experiences are not representations or maps of things but things in their own right. The experience of squareness is a map of certain types of object with four sides. But what is the experience of nastiness? It might be considered to be a representation of a group of things that we associate with some future nasty event, but that still requires the idea of nastiness to go with the events. The idea may be important for survival but that does not explain where it comes from. We use the term emotion to mean a whole cluster of things that belong together in the way indicated above but at the root of the concept of an emotion is something that does not seem to have an outside equivalent. It seems to be a type of relation, in the way that 'belongs to', 'to me' or 'means' are types of relations.

Put another way, whereas thinking of a cell membrane experiencing red or the taste of cheddar is not too far away from the intuitive idea of seeing or tasting 'something', the idea that the feelings of good and bad are patterns in such a membrane seems more difficult. What seems certain is that we cannot escape from the idea that the feelings of good and bad are properties of the physical stuff of cells, which is like any other physical stuff except in that it may be in a particular pattern. That pattern may be complicated, but on the other hand good and bad seem to be the sort of things that are not complicated; they seem to be very simple.

Although I am far from convinced, I wonder whether certain experiences like emotions, the ideas of good and bad, love and hope are things which exist both internally and externally because they are fundamental relations between things. I can see pitfalls for anyone trying to guess what their physical basis might be but there are two things that tickle my interest.

As I have mentioned, Jaak Panksepp (2002) has emphasised the fact that the cells around the cerebral aqueduct in the mid brain, the PAG, have a claim to being the seat of 'full story' awareness. As I understand it, in this area emotion is associated particularly with outpourings of chemicals into the fluid space around cells that can affect cells on a general basis rather than simply at single synapses. There is thus a suggestion that emotions may be associated with very general changes in the electrical activity of cell membranes rather than isolated changes at specific synapses. In some ways it seems reasonable to see emotional feelings as having the job of changing the threshold of response to sensations rather than altering the pattern of sensations that is most appropriate to respond to. Thus in a good mood you might greet the postman and in a bad mood you might not, but recognising the postman would be the same.

The other point is that things like good and bad seem to have something to do with order and disorder, stability and instability, persistence and disintegration. At least in a primitive organism one might expect a fairly simple match between these things. For a field, order and stability imply the existence of a dominant or resonant mode of oscillation. In many systems at the quantum level there is a natural tendency for field quanta to take up the same state – to harmonise if you like – that is offset by the chaos-inducing effects of heat. It would be so simple if happiness were resonance and pain chaos in the cell membrane. Maybe they are, and maybe the chemicals bathing cells in the PAG help them along, but I am not going to be the person to stick his neck out that far. There are some fairly obvious uncertainties here. Happiness tends not to lead to action but pain does, yet in a musical instrument resonance leads to action and chaos does not.

At least relating good, bad, happiness and pain to basic physical states seems more promising than trying to hitch these things to patterns of message-passing between cells. I find it very difficult to understand why a certain sequence of signals in a certain pattern should hurt.

It is easy to say that what one person calls good is not necessarily what another person calls good and so good cannot be anything absolute. However, the concept of good functions as the same relation between other things in both people's experience. I think we can still expect the sense of good to be based on a common basic phenomenon, even if George W Bush illustrates the fact that it is as firmly attached to other concepts in our brains as sticky labels on supermarket fruit. Perhaps our selfish genes have built a structure that harnesses the relation of goodness for their own purposes. That raises the possibility that sometimes something is perceived as good not because it is of any advantage but because it is intrinsically 'harmonious'. What about J.S. Bach? What possible use is it to like the Brandenburg Concertos more than a sonata by Hummel. Was Bach tapping in to neuronal force fields simply enjoying a good tune? Or was he just very clever at mimicking the rhythms and tones of things like human speech in a way too subtle for us to notice? This is dangerous territory but why not? The neuroscientists tell fairy tales, so they cannot complain.

Throwing out homunculi with the bathwater

A common feature of fashions in science is the spread of a 'knockdown argument' that shows that what people thought last year is ridiculous and that anyone who still thinks it is just yesterday. The homunculus concept is widely believed to have been debunked by the demonstration that it requires an infinite Russian doll regress of further homunculi inside homunculi. If there is a little man inside my head repeating the seeing of my eyes with some further 'internal seeing' you solve nothing, you need another little man inside the little man.

However, this debunking only applies to a very crude view of the homunculus, probably based on a rather patronising interpretation of the views of seventeenth century philosophers. In a more sophisticated reading there is no repetition of seeing because the receiving of the view we experience only occurs inside; there is no seeing in the eye. Moreover, if the structure inside that sees is receiving information in its own domain, rather than from outside itself, there is no regress.

The cellular homunculi that I am proposing operate in a way very different from either a videocamera or, for that matter, the seventeenth century homunculus. Patterns of information enter their domains and the meaning of the pattern lies not in individual signals or symbols coming in but in the nature of the pattern itself – in its multitude of interrelationships. This pattern of relationships means something to a fundamental physical element in the domain in the same way that a crystal means something to an x-ray beam passing through it. This meaning seems to be the outside world, but it can only be a very neat match to certain features of the outside world, generated by the information processing system that is a brain. In a sense it is not even a map because maps have parts and an experienced pattern should have no parts in this sense.

thirteen

Understanding

Wherever the building of hierarchies of sensations, alongside perceptions, alongside language, occurs in the brain, it might be expected to have pitfalls. The mysterious understanding that Penrose (1994) and Searle (1997) talk about may have to do with the power of layering relations on to relations. It may be powerful but it may also be precarious. If the way we build up layers of relations is on the basis of 'what matches best with what we already have' (who calls Bingo first) which seems likely, then often that match will be valuable but sometimes it may be flawed. Understanding seems to be a situation where patterns are stacked up so that running a question through the system gets the right answer every time. Misunderstanding is when you seem to get the right answer much of the time, but there are times when everything goes horribly wrong.

Very often it is easy to see how misunderstanding works because it seems to be the attachment of a word to the wrong pattern of things or events. When a general says he needs more troops for the war against terror it is easy to see that he misunderstands what this war is (or he wants others to misunderstand). The war against terror is fought by setting a political example and by airport security checks. Stem cell research seems exciting until you realise that nobody knows what they mean by a stem cell, and even if they did it would be unlikely to regenerate an organ. A 'thriving economy' that creates maximum waste and destruction seems to be the goal of our society. For every example of the human brain pulling off a feat of Penrose's non-computable thought or Searle's true understanding there is an example of an error a computer would never make.

One must assume that the acquisition of understanding will be riddled with false starts – how could it be otherwise if it works on a best-fit basis? Yet when I recently asked interview candidates for medical school places to give an example of something they realised they had misunderstood, but had then come to understand, there were few convincing answers. I find that worrying because I do not consider one of my tutorials successful unless in an hour I have uncovered at least half a dozen major misunderstandings. On a good day I find a big one of my own.

There is an extension to this issue. Because the link between the 'story' and action is not one to one in my viewpoint there is perhaps greater scope for misinterpretations of links between the two getting built in to the future story, even if democracy might iron things out most of the time. Moreover, getting a useful link at all may depend on there being a cell that recognises a pattern that is, for instance, genuinely a match for a word. Most people will recognise blue, but it is not so clear-cut for faces, music or patterns of abstract ideas. A person with perfect pitch can say 'I hear A-flat'. Another person may have, presumably, the same experience of a musical sound but has no way of knowing what the note is. There are situations where we think we are recognising a familiar face in one context when in fact the familiarity derives from another face known from a different context. In brain diseases such as stroke and dementia muddling up of the process of talking about experience is commonplace.

But the humbling (and maybe wicked) thought is that the ability to talk about the story that is in our awareness is, because of its indirect and vulnerable mechanics, something which humans have become quite good at doing but are by no means fully proficient at. It may be an evolutionary development on the brink of becoming a reliable function, but not yet a reliable function. Within the population it may be something that varies considerably for different types of experience. Some people are good at describing what they see and others what they hear despite the fact that they have equally good eyes and ears.

Maybe this impinges on the puzzle as to why some people will have followed my arguments about the problem of having access to patterns from the start and others, if they have even bothered to read this far, cannot see the point. Maybe it has something to do with the fact that some philosophers like Russell, Searle and Chalmers say they see the experiential part of the mind as so mysterious and others like Dennett say they see it as a non-issue. Perhaps the precarious link between experience and language is at the root of the issue of understanding all sorts of things. Perhaps because it is so precarious abilities to understand different sorts of idea are so variable and in some people missing!

And none of us can consider ourselves immune to this problem. It might be easy to say she is really sharp but he is plain dim. But the extraordinary thing about the debate on the mind-body problem is that even the sharpest minds seem to have blind spots. Take David Chalmers and John Searle, both of whom, in my opinion, have made a major contribution to the field. In one paragraph you will find them cutting through the undergrowth of misunderstanding with an intellectual machete in the most impressive way. In the next paragraph you find them planting a strangler fig of an idea that ensures that no coherent thought can pass further. The result is being permanently intertwined in an argument from which there is no escape. Or at least so it seems to me.

Horace Barlow mentioned to me the interesting point that being able to talk about our understanding is not necessarily an unmixed biological advantage. There are reasons not to tell everyone else how to find the food or how to be The Prince. To be Machiavellian you should not be Machiavelli. Maybe the cleverest bits of our brains are not linked to the perceptions we can discuss for good reasons. Perhaps our best understanding is not available in words or numbers.

I have an increasingly strong suspicion that we overestimate our ability to understand things. New human understanding comes rarely and seems outside the range of even the majority of scientists. Einstein developed his theories of relativity not because he had done any experiments but because he was the only one, of many thousands of people, to see that there was only one way of understanding information that had been around for a while.

I find myself echoing David Hume (1999) again. For him, what we call reason is little more than a matter of familiarity; something looks like something we have seen before so we think it works the same way. I am not sure that I would be quite as dismissive of understanding as Hume, but I certainly share some of his scepticism.

At the beginning of the book, and at the beginning of my current journey into the realm of thought, I spent some time focusing on the ways in which brains seem to do things that it might be difficult for computers to do. I was impressed by Roger Penrose's arguments. However, as several people have pointed out, it is not clear that even if our brains do non-computable things these involve awareness. Although understanding seems to imply awareness, and that is why it seems barmy to say that John Searle's Chinese Room understands, understanding seems to arrive in our awareness having been generated unconsciously. I am aware that I understand that adding two even numbers always gives an even number, but I have no awareness of what the mechanism of that understanding is or what lies behind my sense of confidence that I understand.

There seems to be a fallacy in several attempts to analyse the way our brains work, which assumes that thinking occurs as one act in the brain at a time. That is, if I am doing a sum, it is assumed that this involves one process, and since that process is in my awareness the awareness and the mechanics of adding are functionally matched. But we know that the brain is not like this; it delegates jobs to different parts. It is like an old newspaper headquarters with editorial rooms (in the mid brain?), journalists in offices (cranial nerve nuclei?), typesetters (cerebellum?) and photograph archivists (hippocampus?) all working separately and then feeding back to the finished product. That is why the concept of a single entity 'I' both doing and understanding something is problematic.

What does seem to me likely is that the understanding that Penrose and Searle are interested in involves the relating of patterns in the way I have discussed, using some internal language which, at least in part, reflects the rules that Chomsky and others have tried to unearth. Since we are aware of at least some of these patterns and relations there is a link between the mechanism of understanding and awareness, but, as I have covered in a previous chapter, the causal role of awareness is a subtle matter.

The fallacy of thinking as a single action is, I suspect, at the root of attempts to find a new mechanism for thought over and above the mundane chain of H-H waves we know about. William of Ockham's law of parsimony of explanations tells us that we should not look for another site of quantum computation in addition to H-H wave interaction unless we have to. My feeling is that what the 'extra element' people are looking for may just be the end result of all the other millions of interactions going on in the brain at the same time, which, as I shall allude to later will, at least in one sense, not be algorithmic in the way that a computer is.

Nevertheless, I am still intrigued by the idea that wave-based processes, whether they be H-H wave interactions or processes involving another type of 'listening' wave, may have something to do with the 'non-computability' that Penrose (1994) attributes to brain behaviour. Processes involving interactions between observing elements and complex patterns as *single indivisible things* might provide a basis for both 'meaning' and 'understanding' through their internal richness. If this richness allowed complex laws of relations, relevant both to language and to the mathematical proofs Penrose is interested in, to be an integral part of thinking, then understanding would be much less surprising than in a computer that simply adds or subtracts one 1 from another. In philosophical terms, if our brains make use of the complex rules of the universe to think then they might be expected to have a chance to understand those rules in the way that a photon seems to understand the rules of Schrödinger's equation – because it is the universe at work.

The 'feet on the ground' rebuttal of this suggestion is that if what our brains *do* could be entirely explained by electrical potentials adding up in the way they would in a wire cable then it is unnecessary to invoke the more interesting properties of waves. It may be fanciful to link the 'understanding' of fundamental waves and fields to anything a human being says. And that may be the end of the matter. Understanding may just be a convincing fiction laid on by our inside storytelling machinery. We may have no ultimate way of telling whether we really understand something or whether what we think we understand is merely the courtiers assuring us that the imperial gown that to us is invisible and untouchable is of the greatest beauty to all around.

But, as I discussed earlier, there are least some reasons for thinking that the interactions of a listening wave in the cell membrane may affect the world. The effects may be immediate or through the longer-term phenomenon of Hebbian reinforcement. If our awareness is really all about listening waves that control memory and learning then maybe understanding does have something to do with waves that have a meaningful quantum level description even if their effects have a perfectly adequate corresponding classical level description.

When we come to understand something it is often as if a tiny turn of a telescope focus ring has, in half a second, made everything in view have a new interpretation. Moreover, it can stay like that forever. Understanding, as part of learning, seems to be not a computation but a rearrangement within the brain that makes all computations thereafter flow in a smoother, simpler path. Some process with a time scale of a second or less alters the way cells respond and another process, with a time scale of hours or days makes this permanent. This is not adding up of post-synaptic potentials, but the neural plasticity that allows us to learn. I just wonder whether a listening force field in the membrane has the ability to 're-tune' the relations between synapses so that a pattern that generates continued positive feedback becomes fixed in place.

As far as we know Hebbian reinforcement is the process that converts us from dribbling infants into intelligent personalities. It works on the basis of 'custom or habit' that David Hume (1999) claimed as the basis of human understanding. It would be nice to think that awareness had some link to it. Readers may prefer to remain sceptical about a relationship between fundamental wave properties and understanding. Nevertheless, there are some deeply puzzling things about our awareness that seem to need a metaphysical link something like this at the end of the day. Neither the physicists nor the philosophers can claim to have all the answers.

fourteen

A (Very Short) Bit About -isms

All ideas about thinking tend to get packaged into an -ism of some sort. What sort of -ism does my suggestion fall under, reductionism, functionalism, monism, dualism, indirect realism, physicalism, monadism, Russellism, Petrovism…? When drafting this book I planned a chapter on which -isms are better than others; a common pastime of philosophers. But having read more I now know better; it does not matter half a peanut.

The biggest debate seems to be between physicalism (or materialism) and dualism. Yet as far as I can see physics is rampantly dualistic. Any object can be described both as an aggregate of matter particles or as a force field, one taking up space, the other not. Everything is either an observer or an observable. Observers have no definable extent in space and cannot be described by the rules of observables. Even observables obey two sets of rules, one for when you are not observing and one for when you are. Descartes was just scratching the surface.

The point is that you can make these words mean whatever you like. -Isms are fudged ideas used by professional hot air merchants. Every -ism can be made to be right or to be wrong. They are very often used to sound important and sneer from a position of what is in reality ignorance. A few people use them with great care and skill but they should know better than to be bothered with them in the first place. Science does not need -isms. Art does not need -isms. Metaphysics and philosophy do not need -isms. They need clear ideas.

fifteen

Reprise: Why Cells Will Do, But Not Brains

A reprise seems in order, partly to take stock and partly because it has been so much of my experience struggling with this idea; time and time again I ask myself the same questions and time and time again the inescapability of the conclusions returns. Not only is a single global awareness in the brain incompatible with anything we know about the world, it has no logical basis. If you tried to get a computer or an emotionless life form like Mr Spock of Star Trek to deduce the locus of awareness in a human brain from the laws of physics and the available evidence they would discount a global awareness immediately. An aware cell seems odd, but it seems to work.

The number of messages passing across synapses in our brain is enormous. Imagine every man, woman and child on the planet, each playing a different piano piece by Franz Liszt simultaneously and you will have a rough idea of the amount going on in a brain. Our awareness is nothing like as complicated as that. It is associated with a tiny proportion of the signals in a brain. The only question is how tiny.

Why are brains so busy? Quite a lot of work may be needed to get information ready for one awareness, but that much? The obvious suggestion is that the information coming in from the outside and from our memories is being passed round in multiple copies to a very large number of neurones to see if it fits with something that has happened before. We can be fairly sure that this is true because every signal from a cell is passed to hundreds or thousands of other cells.

The most common logical objection to my suggestion is that it is impossible for enough elements of information to reach a cell to explain our awareness. However, a million signals a second is not bad and there is nowhere in the brain where more elements arrive together.

The neural net people say that it is not the neurones that integrate and process information so much as nets of cells, but integration only occurs in cells. The value of having a single brain is to allow lots of different incoming elements of information to be brought together. Until they are actually in one place that has not happened.

To suggest that awareness belongs to a group of cells is to say that cells have telepathy. Why do serious neuroscientists accept this idea while they scoff philosophy or metaphysics. As James implied, to scoff at metaphysics is just to bottle out when the going gets tough.

There might be a kinder way of putting it. Maybe the majority of people in science simply do not see the bigger picture of what is logically possible because their brains are not set up to respond to the right patterns. The ability to think through an argument may depend on the answers to questions already being set up in our brains. This is how the immune system works, the 'answers' have to be in place before the 'questions' arrive. Perhaps a brain is set up to answer lots of questions but cannot cover them all.

Hardly a kinder suggestion maybe, and I suspect another factor may be as important, since the impossibility of two separate things sharing one experience does not seem difficult to grasp. The immune system's T cells refuse to respond to any protein if it presented to them as 'self' at an early stage. Maybe our sense of self makes certain ideas off limits in the same way. I certainly seem to be off limits; in many ways I have reintroduced all the things considered most stupid and old fashioned; homunculi, panpsychism, dualism, second transduction and more. Interestingly, these are all the things that engender an emotional, as much as a rational, response in mind/brain studies.

Maybe it is I that cannot see the arguments on the other side. I am happy to accept that I may be one of the people who cannot understand. However, I would like to hear the arguments against awareness in single cells. It is not that I cannot follow such arguments, I have just yet to find them written down, or spoken of. All one neuroscientist could say to me was that 'I *know* that I have only one consciousness': end of discussion. We need to get beyond that level.

But yet I could still have missed the point. Maybe my brain can only put things together in certain ways. Perhaps I have the same limitations as Leibniz! Leibniz was capable of some impressive thinking, but perhaps it is of a sort with limitations. Perhaps different human beings will always understand the universe in different ways, each valid for that particular mind. Yet this is not the way science tends to go. The mysteries of the past tend to be seen as having one solution. Whether or not awareness is a property of individual cells seems to me to be a question which must have an answer: yes or no.

Some would claim that all that matters is the 'function' of the brain and that it will carry an awareness that matches that function. Function is a weasel word that can mean purpose, effect or action. The function of the heart can be to keep us alive, to circulate the blood or just to pump. To say that something has awareness because of its purpose or effect is highly dubious. A cash machine is unlikely to have the awareness of a clerk. The idea that when two things have the same effect they have the same things going on inside is one of the biggest sources of wrong ideas in biology.

If function means action, that might make sense, especially if we use David Chalmers's idea of 'sufficiently fine-grained function'. But if two things have the same action in the tiniest detail they must be pretty much the same thing. Chalmers (1995) seems to equate fine grain with keeping the logic the same at the network level. He has made use of an elegant thought experiment based on imagining what would happen if cells in a brain were replaced one by one by silicon chips with exactly the same 'function', which in this case means action for networks but effect for cells. The actions of a silicon chip are not those of a neurone. Assuming consciousness arises at the network level Chalmers asks: after how many of ten billion replacements would you stop being conscious? However, if each cell is aware separately the paradox evaporates. Each time you replace a cell you lose one awareness.

But there is the bigger problem; awareness does not match the function of a net of nerves. Awareness is not about signals duplicated a thousand times moving around, of circuits, diverging, converging and re-entrant. If a cell receiving a signal from a red sensitive cone in the retina sends a signal out to 1000 cells how is the function of those 1000 signals 'red dot'. Surely the function will be 1000 different questions, all of which will be unconscious; 'is this part of a tomato?', 'is this part of uncle Henry's nose?', 'is this part of something you have never seen before?'. The awareness may be of a Greek salad. Awareness is just about access to patterns of information; what we know happens in a single cell. One hand cannot clap. Its nature does not match clapping. Similarly, a net of nerves does not match the nature of awareness.

People may find it hard to believe that scientists might fight over ideas simply because most cannot get their head round them. Scientists are supposed to know what they are talking about. The reality is that in biological science experts who do not understand the basics of their subject are two a penny. That is what makes me think that I am entitled to doubt the wisdom of my fellow professors. Perhaps the advantage of studying the soul rather than the immune system is that at least one or two people can see that sloppy thinking is going on.

The one thing I must agree on is that if awareness is a property of individual cells we still have a daunting problem in trying to understand how it could possibly be like what it is like. However, I think it could possibly be like that. I do not think consciousness in a net of cells can possibly be like anything.

If all nerve cells are aware separately it might seem odd that each awareness should think it controls all our behaviour. But does it? 'Unconscious' thought, often seems very clever, as when waking with the solution to a problem 'one' was not aware of considering. Introspection suggests that most of our actions are decided on 'somewhere else'; words appear for us when we want to talk, we find ourselves laughing, and even having made crucial life decisions without any sense of how. At least, the awareness(es) thinking the higher thoughts put on this page feel(s) that way. Why do we not know about these 'unconscious awarenesses'? We can only ever *know* of one awareness. We infer the existence of other awarenesses in other people because their behaviour mirrors the story we hear in ourselves.

Some aspects of mental illness also hint that the 'story of me' is an artificial and fragile thing. In the mentally ill brain the story that there is one awareness in charge often fails. In depression there is often a sense that the person is not just worthless, but 'not me'. In schizophrenia others may appear to be talking inside one's head. I would not say that having an awareness in each cell explains these things but it might make it easier to see how they might be explained.

A problem for most theories of awareness is that they require some nerve activity to be sentient and other activity insentient and there is no hint as to why that should be. Steven Goldberg (2006) addresses this with his idea of 'Jonah minds'. Jonah in the whale has a mind but it is not the whale's mind and only indirectly affects the way the whale behaves. In a sense Steven Sevush and I merely take this idea to the logical conclusion. If each cell is sentient, the problem is not there.

My case is as follows: We have a brain, in which thinking occurs. It has always been unclear to me what 'mind' is and my view would probably make it so unclear that it is best avoided. Thinking involves the creation of an inside story, relayed as an ever-changing kaleidoscopic pattern to millions of separate cellular beings. To these fairly uncontroversial facts I would add that each cell is a sentient 'listener' to the story, aware of the pattern it receives, and that this is the only sort of awareness there is in our heads. The detail of how things work is difficult. Sometimes it seems as if it may fit together elegantly. At other times I fear the whole thing will fall like a house of cards. But it would fall for a single awareness in a brain just as much.

Only by Christmas 2004 did I discover an irony in the term consciousness, a word that I am increasingly ambivalent about, because it has too many meanings. Its original meaning was 'together knowingness', as in public consciousness; the shared knowledge of the people. The word implied a single story common to many listeners. I suspect the consciousness in my head is exactly that. So I could say that I do have one consciousness, but that would confuse everyone, because it would not be consciousness in the modern sense of the word. So I will continue to try to keep consciousness out of it.

You will see why I stuck to a story in my head to begin with.

Darwin's theory of evolution by natural selection requires us to believe extraordinarily implausible things, like the first development of eyes and wings *for no reason at all at the time*. Eyes were used for seeing and wings for flying because they happened to be available. They were not made for the purpose. How implausible is that? No experiments have been done or can be done to prove Darwin's theory. Even the journey to the Galapagos was just a way to get the right person into a frame of mind. But Darwin had on his side the fact that not only was natural selection possible and the alternatives impossible without magic, you could not stop it happening if you wanted to.

The idea of an inside story with many listeners is as implausible as natural selection and there may be no experiments to prove it, but the alternatives seem to me as impossible as Darwin's.

IV:
What Does It All Mean?

sixteen

How Should We See Ourselves?

Behold! Human beings in an underground den... Like ourselves... they see only their own shadows, or the shadows of one another, which the fire throws on the opposite wall of the cave.

Plato; The Republic, bk. 7, 515b

The idea of a single story of self with many listeners, each able to tip the balance of future action in its own tiny way may well leave people confused. It should do, because it requires a complete rethink of our view of ourselves. Having embraced the idea I found myself needing to reconcile it with what faced me waking up in the morning; real life. These are some of my thoughts.

Not all of what I am saying is new. I do not need to feel responsible for doing some terrible damage to the human soul because Daniel Dennett (1991) has already done that. For him a person is just a story, kept as lots of little snippets or 'multiple drafts' around the brain with *no listeners at all*. Dennett has already cut the head off the Greene Knight and kept him talking. I am letting him feel as well, and in glorious multiplicity. I think that both the doing 'I' and the listening 'me' are real but the catch is that I do not think that they are the same thing; whereas the doing of the brain is built up from the doing of all its cells the listening remains private to each cell.

How can we conceive of ourselves in this curious divided way? Having pondered over this for many months the conclusion I always come back to is that we should not expect anything to *appear* to be different from what we are used to. The viewpoint I am proposing does not change the way we think the brain *works* in any critical sense. Whether they be from

Crick (1991), Penrose (1994), Hameroff (1994), McFadden (2002), Pockett (2002), Pribram (1991), or Vitiello (2001), other suggestions for how consciousness arises all tend to imply new ways for cells to talk to each other which as far as we know do not exist. My suggestion relies on cells communicating only the way we know they do.

One might think that a cell might come to think 'I don't like the way some of the other cells in my person are behaving'. But an individual cell will have no concept of its own identity because a concept of identity is part and parcel of the story being generated by the action of many cells at once. The information coming in to these is not about the environment of the cell as a cell but about the environment of the person. Maybe in evolving from an amoeba, that does have information about its environment, to a neurone, the neurone has lost its own story by putting its senses to communal uses, rather as Nick Humphrey (1992) would suggest.

Each cell will respond in a particular way to the story of the person but there is no reason to think that the cell has any knowledge of whether it has responded and whether its response had any effect. A cell will never get a meaningful message 'I am the cell that just made that decision', and there would be no point in such a message.

Selves, souls and SESMETs

It might be time at this point for me to decide what I think self and soul mean. For a while I thought it would be convenient to call the story that goes on in a head the self. This would fit nicely with the concept of a story of self, developed by various theorists. It would leave the option of calling the subjective being, which I see as a cell, a soul. This would be the soul as Leibniz meant it, although he saw there being one 'supersoul' for each human being.

The difficulty with all this is that the word self has also been firmly attached to the subject, particularly by Galen Strawson (1999). There may be a way round this in that Strawson has invented the SESMET (Subject of Experience that is a Single MEntal Thing), which would fit with my cellular subject. This might relieve the self of its duties but I still see problems.

I think I will steer clear of both self and soul. The story aspect I can call the personal story or just the inside story. The cellular subject is just a neurone. Francis Crick (1994) said that we are nothing but a bunch of neurones; indeed, a *bunch* of neurones.

This may leave some people bereft. Where did my soul go? I am tempted to say 'whoever said you had a soul in the first place?'; people who want you to drop coins on to their silver salvers mostly. But maybe something can be rescued. I am afraid that I have never considered it likely that there is any sort of soul that survives after the death of the body. Having been a doctor for thirty years I have often seen bodies that may still smile, even walk and talk but which are inhabited by only a remnant of a person. Souls in the sense that most people consider them are more fragile than bodies these days. If the soul is something with an experience, like Strawson's SESMET, I fear it may be a force field that lasts a fraction of a second. If it is the cell membrane that hosts these experiences, even if they could in theory be about different people, and according to some claims sometimes are, it probably lasts most of a lifetime but no more.

But if there is something precious that might be called a soul then surely it is not a cell, but the story of the person. Granted, the story would not be the subject. It seems that it needs listeners. But that is not so bad. It is not so far from our intuition or the position of the great writers of the past to say that the precious centre of a person is the story. When a person fades with age we see a story with no new chapters, a story that begins to repeat and lose interest. When one meets a teenager blind and deaf from birth a wall of useless pity comes down until one sees the child writing her story with her finger on her father's thigh knowing that he will write your words back on hers and keep her story full. How important the story is to us is made clear by hungry fingers eating Braille from a wheelchair in a hospital corridor. This is where Dennett is very right.

And maybe stories do go on after death. Perhaps immortality is our story going on to the next generation. Playing Chopin or Cole Porter makes one feel that these were people who knew how to go on beyond their own listeners in style. But, as a friend who had lost a much-loved father pointed out, the story taken up by listeners in other bodies is not the original story. It may have some of the same patterns but the inside story is lost at death. Maybe that is just a reminder to get on and create as rich a story as one can before it fades. It still leaves me with the contentment of thinking that what matters is what has always seemed to matter, the story that we are.

If we want to make ethical decisions about how we treat animals or unborn embryos or, for that matter, computers, the answer must be that we judge what sort of story we are addressing. It does not really matter how many listeners there are, but whether or not there is a story worth listening to, whether it is rich. As John Donne pointed out, you also want to know how much that story becomes part of your story, how much the bell tolls for thee.

How should we see other people?

What should one make of other people? Should one look them in the eyes and talk to them as if they are a million listeners? I think not. The concern raised in the subtitle of the book can be laid to rest. There is only one person in Siân's head and one in Beth's head. Talking to them as if there are many of them is as foolish as it feels. If I talk to a class of students I know I am addressing many stories. Some are interested (maybe), some worried they might be asked a question they do not know the answer to and some want lunch. Talking to one person, talking to six people and talking to a hundred people are different activities. And talking to one person, who is a single story with many listeners is still talking to one person.

Nothing is new. We should see ourselves as we do; but probably the way that writers and dramatists like Shakespeare, Wilde, Hemingway or Mike Leigh do, rather than the way psychologists do. The constant flux and tension within the personal story, which no doubt reflects the complex connections in our brains, seems to be illustrated much more directly by literature than by science. Just as arithmetic may seem a counterintuitive clumsy contrivance in comparison to our innate use of space and time, so psychology often seems a clumsy parody of how to understand each other.

Perhaps psychology just has to find the right tools. Maybe it can show that Shakespeare knew certain basic rules that can be translated into scientific language. On the other hand the building of an inside story may be one of those bits of biology that derives its success from having no rules, like the evolution of antibodies and genes. There may be as many right ways to look at our inside stories as there are cells in a brain because having so many ways of seeing things is how the story works. Maybe you explain stories best with stories. Nevertheless, if we understood the interactions in the brain that form the elements of language perhaps we would begin to understand how this can be.

Evolution at its most Machiavellian?

Many people have wondered how human thinking evolved. It is often said that we are the only animal with a true sense of our identity. A number of theories revolve around the idea that having a story of 'self' and maybe having a big brain has something to do with complicated social interactions, but I find these ideas a bit difficult to match up with a bee, a cow, a wildcat, a dog, a baboon, a bonobo, an eagle, a hummingbird and a crow. It seems to me that even if not communicated to others there must be some version of our sense of self in many if not all animals. It would help to make sure you do not eat your own leg.

But the irony that I am suggesting is that this sense of a self is a trick, an illusion, a clever deliberate mistake on the part of evolution. It is a way for a colony of one-celled beings to think that they are one being. Until now the illusion has been almost unchallenged, although it may have been tinkered with in the nineteenth century. What I am suggesting is a frontal assault; our sense of self is a complete cheat.

Nick Humphrey, in *A History of the Mind* (1992) has written with insight on the evolution of the human mind as an internalisation of the interaction of a primitive organism with the environment through its surface. He sees it as a way for a complex organism to maintain dialogue with the world with the advantage of a central site for integrating patterns of stimuli not available to a simpler organism. I follow much of Humphrey's view but would simply add that this is not a strategy for a single being but a strategy for a colony that survives if every member survives, a colony of altruistic individuals. The really clever thing for a colony of organisms like this to do is to think that they are one being. Although verbal language is not there, this should apply in some degree to all multicellular animals. In cold-blooded 'functionalist' information processing terms it is probably fair to say that anything with a single central nervous network works on the basis that it is a single being. It is just that the ability to sense this, ponder on it or talk about it presumably increases as you get nearer to man, whether or not you consider the change from ape to man to be the most important.

Free will

Many people come to the mind/body problem wanting to find a place for free will. I remember as a boy in the 1960's being intrigued by the popular idea that the randomness of quantum theory allowed room for a *deus ex machina* intervention of free will that would affect the future in a non-deterministic way. However, my observation of the human condition as an adult, particularly as a doctor, seeing how people behave in the context of major life events, leaves me thinking that there is no need to suggest that there is this sort of free will. There is no evidence it exists. We behave the way our genes have programmed responses to the environment, perhaps helped by the random generation of signals from within, as the immune system does it. There would be no reason for us to do anything else, unless God told us to, and that would not be our free will.

In fact I find it hard, yet again, not to leave it all to Spinoza, as quoted by Roger Scruton (2001): 'Men are mistaken in thinking themselves free; and this opinion depends on this alone, that they are conscious of their actions and ignorant of the causes by which they are determined. This, therefore, is their idea of liberty, that they should know no causes of their actions. For when they say that human actions depend on the will, these are words of which they have no idea. For none of them know what is will and how it moves the body: those who boast otherwise and feign dwellings and habituations of the soul, provoke either laughter or disgust.'

My feeling is that free will arose in the past as a concept of being able, as an individual, to act according to one's desires rather than laws sold as coming from God, but actually coming from social and political pressures. Nobody was ever free of God, or the universal rules, merely free of rules made by humans, as in the French Revolution.

Neither standard neurology nor the idea of awareness in single cells makes much sense of free will. If free will is to do with what output you get from an input it has to be a property of individual cells because they are where input leads to output. The only 'choices' made in the brain are individual cells choosing whether or not to fire action potentials. Nothing else in the brain can determine anything. An individual cell seems unlikely to have an opinion of its own what to do other than in the sense that its physical structure determines its responses. I cannot see that we can expect to 'buck the system' and act against the laws of physics. Neuroscience gives no support to the idea that there is any central, aware, decision-making soul. Our intuitive impressions, and Libet's experiments, indicate that most of our decisions get made without us being aware of the process. The idea that 'it was I of my own free will that did it' is just part of the illusion of being a single agent.

If we have no free will it might seem that our actions are laid down at the beginning of time, that we just act out the algorithm of fate. However, this may be as misleading an idea as that of free will. People talk about the random nature of quantum mechanics and others doubt the relevance to biology, but it seems to me that uncertainty is built in to any description of the universe for much deeper reasons. And Spinoza makes the point that our actions are indeed our own because they belong to that unpredictable bit of the universe that is us (Scruton, 2001). There is no other global thing that dictates what we do, even if there is a global set of rules that we follow. We may not have free will but it is our little corner of the universe that is responsible for its own actions, not the rest of the universe. It is perfectly reasonable to punish criminals.

At the beginning of this quest I was impressed by Roger Penrose's account of how we seem to be able to do more than a computer following an algorithm (Penrose, 1994). I remain uncertain how to interpret our apparent ability to 'know' things that Gödel suggested we could never know for sure. But, while arguing over algorithms and fate with Miguel Nadal it struck me that the fact that we do not seem to carry out algorithms is not surprising.

An algorithm is a set of instructions that dictates events in a fixed sequence. This applies to the working of a standard computer in which the important events occur in strict series, rather than several things happening at once. Computers with parallel processing can have several things happening at once but presumably they still have rules for the order in which interactions between branches of the algorithm occur. Conway's 'Life Game' is an example of an algorithm that mimics the evolution of a complicated system like a living organism. Quite simple rules applied to 'cells' in a two-dimensional matrix can make things grow, divide, glide about and all sorts of interesting things. The rules dictate that each of several things change during any one 'time step'. However, while the rules are being applied and the changes calculated the system is frozen in time. The changes are then all implemented instantaneously. An alternative way of doing things would be to apply the rules to each element in turn, visiting each element once for each 'time step'.

Real things do not work like this. Interestingly, although it might not seem to matter whether you assess the changes in all the elements at one go while freezing the system in time, or you assess each one in turn, for a complicated system these two approach rapidly give very different results. You cannot convert a complex system into an algorithm without distorting the sequence of events slightly. The trouble is that for very complex systems like living things even a slight distortion may totally change events quite quickly. An algorithm that alters one event in a trillion will have an effect on a human cell in less than a nanosecond.

In a real life system lots of things are changing during a 'time step' and at the most detailed level there is no way of determining which change comes before which. Leibniz realised that at the smallest level you have to have something like Heisenberg's Uncertainty Principle because certainty at the finest grain is self-contradictory. The universe cannot at the smallest level be made up of billiard balls knocking into each other in a way describable by an algorithm. There will always be situations where the difference between the likelihood of each of two possible outcomes (ball A hits B before C or vice versa) is infinitesimal, yet one outcome must prevail. Algorithms can only apply to reasonably big simple things.

As an illustration, if a falcon approaches a flock of pigeons some will be looking slightly to the right and some to the left of the falcon's track. The whole flock will follow the first bird to take off – either right or left. They are no ascertainable rules for which bird takes off first. A brain with ten billion Bingo Ladies working on the basis of who shouts first will be very non-algorithmic because in such a complex system sequence is indeterminable. This is not to do with quantum theory; ordinary classical biological processes are messy enough to ensure it.

Thus, when Leibniz said everything in the universe follows predetermined rules but not a predetermined course he had good reason. Since what a brain does is affected by what it did before, our actions can be seen as a chain of processes which fate did not lay down. Over a lifetime the 'freedom' of the evolving story may play a very large part in generating a unique personality. If that is free will then so be it, although I do not think it is what most people who would like there to be free will would like it to be.

Childhood

A major change in 'being aware of self' or having an inside story seems to occur some time between the ages of nine months and two years, alongside the development of language. This is when awareness of other selves or 'theory of mind' also seems to form. A further and perhaps more interesting change seems to occur at about the age of eleven. At this age children begin to want to develop their individuality. They make considered choices, rather than simply choosing the nicest thing available. These changes raise the question of whether the structure of the inside story changes at specific stages in life and even whether different types of cell come to be the listeners of 'customary' or verbally reportable consciousness. Children under the age of eleven rarely have discussions about their minds and their awareness. Are the relevant cells not ready?

Such questions may be entertaining for long winter's evenings but may also have implications for disorders such as autism, Asperger's syndrome and related states. Is it that in these conditions the role of certain neurones is never fully developed? Perhaps in the brain's society of cells there are some destined to be poets and elders that will explore and solve the bigger questions of life but only after a long period of passive tutelage. Perhaps the storytelling is passed on up from one group of cells to another.

This also raises issues about education. If a society encourages children under the age of 11 to behave as adults, to turn them into consumers, does that mean that this apprenticeship for adulthood is never used? Are we feeding adult stories to cells for which they are not designed? Will the next generation be able to see the world in the new perspectives a changing environment requires? Will everyone be texting their friends when Rome burns this time? Maybe, but I suspect the tensions between forces which drive the building of a story of self have been the same for centuries.

Am I still here?

A final issue about who we are is the worry that if all that is left of our 'souls' are evanescent waves how do we have a sense of continuity. How can I be sure I will wake up tomorrow as the same wave domain? This problem is a non-problem. To think that one existed yesterday is just to inhabit a brain with a memory about previous events in the same brain. 'Was I the same person yesterday?' has no meaning and it really should not worry us.

seventeen

Religion, Politics and the Single Cell

Drifting happily through Camden Town in my fifteen year-old broken down Peugeot, I heard an astounding conversation on the radio. Spokespersons for various religions were discussing the Last Judgement. The Muslim said that it was undeniable that all non-Muslims, including the Christian present, would burn for eternity in Hell. The Christian said that it was equally undeniable that all non-Christians, including the Muslim present would burn for eternity in Hell. Neither seemed particularly bothered about the plight of the other. When pressed they said that it was of course possible that God might make an exception for the other speaker. That is not much help to the other few million adherents to either faith.

Religious people often suggest that other religions are worthy if misguided institutions. Religion, or faith, is seen as inherently good. Yet this conversation made it clear that, if taken seriously, Islam to a Christian must be considered deeply evil and vice versa. A teaching that condemns millions of people to eternal torture must be evil. The hypocrisy is absolute. Only one such religion could be good.

In fact, to the rational person, all such religions, at least in their dogmatic forms, are evil, in that they are based on lies and fear. Why invent hell? Presumably to frighten people into giving you money. I have always thought the teachings of Jesus of Nazareth on personal conduct very reasonable but that has nothing to do with religion. As my father taught me, they form the common sense contract of do as you would be done by. I can accept that science might lead us to an understanding of the universe that allows the word God to be used meaningfully. However, I can see no reason to think God bears any relation to the fabrications of Islam or Christianity. I am even happy to believe that there is already a 'good religion' out there, perhaps the teachings of the Buddha, but certainly not an Abrahamic one.

It may be a vain hope, but if the idea that we are colonies of cells listening to a shared story can be shown to be the only meaningful explanation of our existence there might be a marginally increased chance of freeing the world of the poisonous blackmail of organised religion. Although Darwin's theory of evolution was seen as challenging Christianity it is unlikely that Jesus of Nazareth would have been disturbed by the idea that the Book of Genesis was allegorical. An intelligent Christian has little difficulty absorbing evolution into their faith. If the suggestion I am now making is correct the problem for Christianity and related faiths is altogether more terminal. As far as I can see there is no longer any reason to think that human beings have anything that can be described as a single soul of a sort that might be 'saved'. And of course if we are made in God's image God cannot have a soul either.

What seems to me much more likely is that God is made in our image, and that he is an extension of our illusion of being a single sentient agent. The clever trick of programming of our brains that makes a colony of living beings work to a common advantage through this story of a single self, both for oneself and for fellow animals spills over into the creation of a similar story for the universe. The need to believe in this sort of story is so powerful that even the most rigorous philosophical thinkers often try to build it in somewhere.

You might ask why it matters if people hold crazy beliefs; that is their business. But it is all of our businesses because we cannot go on living by the rules of the past much longer. Neither the planet nor our species can sustain it. We have been brought up in a culture that is under the illusion that Darwin's forces have been suspended for our species. Yet this is only because of an anomaly created by the industrial revolution. Darwin's forces have receded temporarily but will inevitably return, perhaps with the force of a tsunami, in the near future.

Unless people see religion for the nonsense it is we are not going to find a practical solution to our problems.

Some might hope that, like Leibniz, one could still find ways to reconcile a God and God-like souls even with what I am suggesting. However, I see this as clutching at straws from a roof that has blown away. Leibniz believed that the tiniest elements of the universe were souls. He also believed that some souls come to dominate souls nearby, as for the central soul of an animal. He also had a concept of God as a universal soul. His idea that the smallest souls are incorporated into more complex souls is one of the strengths of his writings and I think fits well with the modern notion that individual indivisible matter waves are incorporated into the waves of force fields. He even describes the essence of a soul as an internal force 'like a bent bow'. But I think the extension of this idea to God hits trouble because force fields are patterns that relate to asymmetries within the universe and the universe has nothing to be an asymmetry of.

At a time in human history when the existence of a single human soul seemed a natural assumption and its nature remained a mystery there was some logic in accepting the supernatural. Without a physical explanation the soul can perhaps only be seen as supernatural. It would not have been unreasonable to argue that human souls must be created by some more universal supernatural soul made of the same stuff. The deriding of the supernatural by a science that could do no better has not been that clever, as Penrose's idea of an *Emperor's New Mind* indicates. However, if it becomes clear that each cell is a separate listening 'soul', if the word means anything, and that this can be described in physics, then the supernatural is no longer needed.

Religion often seems to confuse the ineffable and the wonderful. I do not see why it should be the unknowability of the universe that makes it awesome. Even if the suggestions made in this book are wrong, I strongly suspect that our sense of what is unknowable will change radically fairly soon. Certain things will remain unknowable, like the number of atoms in a star. If our awareness of the world arises out of a certain type of wave in a cell carrying a map of space and time we may also never have much information about the way a different sort of wave would perceive space and time. What I think might disappear is the idea that aspects of our world like awareness are beyond our understanding. I see the possibility of a much more complete sense that we know roughly how everything fits together. If God is the ineffable it is about time we got rid of him. Fairly soon there will not be much left of him anyway and if we want to we can worship and be grateful for something real. If we are going to have a God let us have one we can both experience and explain.

The other confusion I see in religion is between the ineffable and the ethical. If you want to enforce a code of behaviour it is a convenient con trick to say bad behaviour will be punished by an unknowable force from which the code emanates. The real problem with Darwin's theory was not that he challenged the truth of the Old Testament. By pointing out the link between behaviour and survival he implied our ethical codes are simply rules emerging from natural selection. They make perfect sense without any God.

Yet, for many people there has remained a gap between the jungle law of survival of the fittest and our internal sense of good and bad. There is a desire for good and bad to be something more fundamental, even if on many occasions they are highjacked by human vested interests and misapplied. I have sympathy with this but, as I have indicated before, however odd it might seem, I think there are reasons to think that good and bad are no more than patterns in the liquid crystal of a cell membrane. They may have something to do with order and disorder at a fundamental level but there is no reason to think that that necessarily translates in any direct way to choices that will guide us to a happy life. The world we live in is much too complicated a place for that.

There are further confusions about the relationship between ethics and life units that may become easier to untangle if it becomes clear the each cell membrane is a separate listener. Cell membranes have quite a good claim to being the ultimate mark of life. Life comes in units and the outer membrane is what makes a unit a unit. We are familiar with the concept of selfish genes but cell membranes are as much self-perpetuating and copying structures as genes. Perhaps they are selfish liquid crystals. In some ways membranes are more actively self-copying than genes. Genes get copied because of a complicated set of 'molecular machines' powered by cell respiration that puts them together. Liquid crystal membranes form bubbles or 'liposomes', as my godfather Alec Bangham christened them, on their own, growing just as other crystals do, because their molecules 'like to be' in rows. As Guiseppe Vitiello (2001) points out, crystals, like life, are by nature things that propagate order in a universe of increasing disorder.

Unlike many other crystals, liquid crystal membranes also tend to break up into packets of a certain size. The origins of life almost certainly lie in this sort of self-copying molecular order, generating packets inside which things can be controlled. DNA came later. One could even look on genes as viruses that have taken over the inside of life packets, maintaining their own existence by helping the life packets to maintain theirs. Life is about packets, about single cells. Animals with many cells are not life packets but colonies of life packets. Similarly, ova and spermatozoa are life packets, even if they only contain half the amount of DNA that other cells do. If the same liquid crystal that defines the packet can act as an indivisible awareness things seem to hang together.

If every cell is aware in some sense it is illogical to suggest that 'life' starts at conception. Everything is cells, and life, before and after conception. There is equally a lack of logic about the idea that life should not be destroyed if it is well enough developed to feel pain or distress. We have no reason to think we know when that might be. Pain and distress may be basic elements of nature. It is difficult to see why they should arise with complexity. Why should you need 976 cells to be sending each other signals to make pain; the idea has no basis whatever. It seems more likely that pain relates to a very basic pattern. And why should an amoeba not benefit from pain as much as us? The purest distress might even be felt by a photon of light. We simply do not know. This might seem to leave us with no ethics, but I would disagree. It does mean, however, that we have to take a rather more sophisticated view of how to weigh up the benefits and costs of our actions.

We need to consider carefully what we want to preserve and nourish and why and what the costs and benefits of various strategies might be in the long term. Perhaps it is stories that are most precious, but to whom; it is not an easy matter.

Sadly, in the twenty first century our system of ethics seems to have become so confused that it is forcing us headlong towards destruction of the planet and maximum human misery. Within developed nations capitalism ensures that rather than protecting common interests and the environment, laws are designed to encourage competition. Yet competition is an obsolete activity since it has no bearing on survival. A billionaire is no more likely to have twenty grandchildren than a single parent on social support. Capitalism depends on growth and there is no real growth in developed nations; the population is static and all usable land is in use. In developing countries the population rises, and misery increases daily from civil war and famine generated by the interference of nations with poorly thought through ideas of morality and ethics.

The key feature of life is not growth but order. It is unlikely that more carbon atoms are bound into living things now than 100 million years ago. Life stopped growing long ago. All it can do now is to increase in the richness of its order. Our main threats are ourselves, and our unsustainable numbers. Put bluntly, I would feel happier and safer in a world of one billion people, still able to enjoy other life forms, rather than a world of ten billion people living amongst a few engineered organisms designed to keep those people alive, and most of them only just. I would like my daughter's children to live in a world full of wonderful animals and birds, in the knowledge that the cells inhabiting the brains of these animals are just as noble and aware as those in our own heads and that *the richness of the human story is the richness of the world we inhabit*.

There are some difficult issues to face up to. As long as capitalism and quasi-religious ethics dominate our culture donations designed to save an African child from starving to death will simply condemn more children in the next generation to the same fate. Perhaps if people begin to consider honestly what their true nature might be some of the unlearned lessons of Darwinism about the way to run our world might finally penetrate.

It may be specious to make the comparison, but the politics of our brains may tell us something about the politics we need for society. When I first indicated my thoughts to the physicist Jack Sarfatti he thought I meant that one cell rules the mind – Stalinist dictatorship he called it. No, for me every cell casts their vote towards the way we behave – true democracy. However, it is not a capitalist democracy of competition. There is no point in one brain cell competing with another, because survival of one brain cell is interdependent on the survival of all brain cells. The brain is a true socialist structure – from each according its ability and to each according to its needs. In a society where smaller will be richer and growth is an anachronism, that might be a fairly good place to start. But only start.

The politics of science

So much for the way science might suggest changes in religion or politics. I will finish this chapter with a few comments on the influence of religion, politics, and science, on science. Much of what I shall say has been said by others, but I am intrigued by the way the problems of science as something shared by a community may be reflected in the way our 'minds' work as communities of cellular beings.

We are used to the idea of religion blocking the advance of science, most clearly in the days from Galileo to Darwin. Religion is now a less powerful brake, even if in Islamic countries at least the female half of the population are often denied access to scientific understanding by religious law and in the United States Christian fundamentalist views seem to be growing. My impression is that hindrance to scientific understanding comes increasingly from within the scientific community itself.

When I first entered medical science in the 1960s big ideas had been cropping up all over the place. Science was flexible and open minded. People talked of, and books discussed, what might explain something rather than what the dogma stated. Nothing was certain, there was the exciting possibility of working basic things out. In contrast, as pointed out by James Le Fanu (1999) in *The Rise and Fall of Modern Medicine*, the last two decades have been full of new technology but very few major new ideas. What people are often unaware of is that science can go backwards as well as forwards. Following the fall of the Roman Empire it went backwards for several centuries.

Understanding is a tender plant. In many areas of medical science my impression is that understanding of the way diseases work has actually been lost since the 1970s. Medical students are ignorant of debates about processes that are still wide open, unaware that we do not even have a clear understanding of why people's feet swell up or why they get short of breath. An explanation may become accepted because it is easy to follow, but the explanation may be self-contradictory or shown to be impossible by historic data. Medical science is treated like a set of Holy Books rather than a dialogue between evidence and hypothesis. Groups of people come to see themselves as the keepers of the Holy Books and talk of something called 'good science' which is really what they and their chums believe in. Scientists have taken over from the inquisition, except that heresy is now punished by unemployment.

Nobody can be blamed for not understanding. It is the journey of science to find problems on the boundary of our abilities to understand if not beyond. It is the duty of scientists to venture into areas where their understanding has only a slim chance of coping. What need not be forgiven is the failure to admit that you do not understand and to preach from a position of incomprehension.

The way sciences shifts forward or back with no final arbiter may mirror what happens in our heads. Just as there is no one person you can turn to in science to ask which theory is right there may be no one arbiter in your head of what you believe or think. All is a consensus. And there is no guarantee of agreement. Just as there are incompatible opinions between people there are often incompatible opinions within one person. Understanding and misunderstanding are much easier to understand if there is no one final arbiter. And this need have nothing to do with awareness in single cells. Again I agree with Dennett in that choices are made in many places at once in our heads, through the electrical events in cells. There is no central choosing station.

Many people will be familiar with a book called *The Secret Garden*, by Frances Hodgson Burnett (1994), about an apparently invalid child who is led to his dead mother's garden by a more adventurous child. The book may seem just a romantic story, a soppy story, but it is also a powerful allegory of something more sinister. The child is not in fact an invalid, he is encouraged to think he is by a nurse who would lose her job if he were to recover. The story is about how we hide knowledge and understanding from others out of self-interest.

Self-interest is usually based on money. Everything is now distorted by commerce. Virtually all funding in biomedical science is directly or indirectly controlled by industrial interests, many of which are not helped by people understanding too much. What about the independence of universities? A silly question, since universities are now run on commercial lines. The main purpose of any research project is to increase revenue flow to the department, through grants and papers for the Research Assessment Exercise = cash. Academic societies are the same. At a recent American College of Rheumatology Annual Scientific Meeting I heard an officer of the College comment on the good relationship between the city of New Orleans and Business. He regarded the meeting as a business venture.

Much like the brain, it would seem, science maintains a confabulation of a single story, a single truth. It achieves this through a thought policing called peer review, again permeated by vested interests, such that 'difficult' ideas are only kept alive by obstinacy. Peer review used to protect people from wasting money on incompetent science. It no longer does that because the journals are happy to sell anything. I often review papers about non-existent phenomena, which get published because they are fashionable. Peer review is now a stamp of approval for those incapable of their own judgement, contrary to the whole point of science.

Peer review generates the illusion of a scientific community with a shared understanding even when no such understanding exists. The brain may rely on constant checking with reality to ensure its understanding at least seems to do what it thinks it ought to do. But if the reality check is replaced by a money check, who knows where things will lead? I guess we have to hope that because the politics of brain cells is co-operative, internal thought policing, being for the common good, may still stumble towards the truth.

And the co-operative spirit is not entirely dead in science. Until now peer review has had a residual role in clarifying scientific writing before it goes to print, but with electronic storage there is no reason why that should not now occur through open dialogue on the Internet. A few brave people are using the net in this way, and a few more in studies of the mind, but not many. There is no money in it.

Philosophers like Kuhn and Popper have frequently told us that science, like understanding, may not work the way we think it does. Unfortunately what the philosophers say often gets garbled. The myth is perpetuated that science is founded on experimental technology, yet most advances in science have little to do with doing experiments the way they have been done before. They have to do with new ways of looking at things, either requiring new sorts of experiment or even no experiments at all. Perhaps the single most enlightening lesson I have ever had was when my physics teacher, James Bauser, told us to work out the surface tension of water with no apparatus or instructions. What at first seems impossible is easy. You remember how big drips from a tap are and with a little simple maths come up with about the right answer. If you want to you can then do it with precision instruments. The point is that real science is not about copying how other people have done things but thinking of a new question and making up how to get an answer. How do we do that? How does a group of Bingo Lady cells get that started? One can only guess.

The peculiar thing about ideas about the self and the mind is that there is no agreed dogma because nobody has been able to find an idea that makes any sense. Instead, people go into little huddles and try not to talk to people who disagree with them. The 'respectable' journals publish none of it unless the author has white hair and a Nobel prize, in which case they may publish, even if the ideas do not add up and are laced with rude comments about other people.

But there are rays of hope. Sometimes one gets a good email. I am optimistic that the reason why nobody can work out what to think in this field has less to do with social and political pressures and more to do with the more benign co-operative politics of the mind itself. If ideas about the self are stuck mostly because the brain wants us to go on thinking we have a single unified soul, we can address that. We only think the Secret Garden is locked; all we have to do is lift the latch.

It may go on suiting many people to see themselves as a single self, but I personally believe there is more out there, there is a garden to be visited. Opening the gate it may genuinely take us beyond the state of the other great apes. For me, the idea that we differ from them in a sense of self looks to be wrong. Other apes may well see themselves as individuals. What may be different about us is that we may be able to see that we are not single selves, yet can still benefit from the great co-operative power that the mechanism that generates a concept of a single self has brought us.

Into a Secret Garden

I come to the end of my story about the stories inside our heads, which itself has become an integral part of the 'story of me' for five years now. Is the story true? Is it a dream, a blind alley like so many others in science or philosophy? Is it time to get back to normal life?

I think I know the answer, even if you might say that I would think that. I think it is near enough to the truth that if it is not true, then any idea that proves it is not true will be as revealing, and as peculiar, as the idea itself, even if it turns this idea upside down. I can no longer take seriously the idea that there is one aware being in my head. The listener, me, is a proud member of a colony that works together well in many ways, even if it has its defects. And at least there is no doubt that the story heard is that of a colony, the colony of cells that is the doing I.

So, if we are stories with many listeners and we become aware and convinced of it, where does that take us? If true this is not just another dry piece of academic biology to file in the textbooks. It changes things in a way such that they will never be quite the same again. What Secret Garden is waiting to be explored if the gateway to self-knowledge, kept closed until now by the forces of Darwin's natural selection, is opened? We shall see, but I like the view in.

I have always been a gardener (maybe a constant gardener for a while). I am attuned to gardens; you may prefer a different analogy but when I think of Francisco de Icaza's line '... *no hay en la vida nada como la peña de ser ciego in Granada.*' (...there is no pain in life greater than to be blind in Granada.) it is difficult not to consider the Generalife Gardens as the place to sit and contemplate the wonder of our awareness having feasted on the shapes and colours of the palace. What so far do I find in my Secret Garden; what might I encourage others to look for?

Perhaps the first thing is a sense of liberation and relief, a sense that everything might fit together and even that art, philosophy and science might again become a seamless whole, with words, pictures and numbers taking their rightful places; a hope that Abrahamic religion might be seen to be the sham it is and the hypocrisy that goes with it might fade. Surely a world that begins to make sense is more wonderful than a world based on a physics that refuses to explain and a psychology that does not know which end to start.

There is also the frisson of the idea that we have the evanescence and plasticity, but also the potential immortality, of stories. Our individual nerve cells may be the individual beings but in themselves they are nothing more than mirrors to a scene. The person is the story, and it is the story that has life in the exciting sense. It is an old lesson; The Tempest, Heart of Darkness, and A Room with a View come to mind; stories about people who are stories. In comparison, David Lodge has some reason to portray scientists as a bunch of self-important charlatans obsessed with minutiae, unaware that their ideas make no sense to anybody (even if they are a nice enough bunch).

All that I can perhaps say at this stage is that the view through the open gate into the Secret Garden is intoxicatingly novel. Even if the conjecture is wrong the fact that we have no reason to think it wrong suggests to me a more open view of our perception. Of the metaphorical contents of my newly discovered garden I would single out three things: parrots, paintings and people.

Parrots

Parrots are symbolic because their true beauty is not as expected. Parrots are often difficult to see in the wild. Their gaudy plumage can be surprisingly hard to pick out in bright tropical sun. When we see a parrot in a cage it is the colours of the feathers that strike us as beautiful. The movements and manner of the bird are those of a clown. In the wild everything is different.

For me, the beauty of parrots is in patterns, but not colour patterns, patterns of movement. One of the most wonderful places we have visited is a lodge in southern Costa Rica called Lapa Rios: rivers (rios) of scarlet macaws (lapas). The most remarkable sight in this corner of rainforest is the flight of the macaws. A hundred or so feet up a red ribbon floats across the sky with a grace of movement impossible to imagine. The tails that in cages look absurdly overlong are even longer in wild birds undamaged by iron bars, but you can see why. Six birds in a row make a perfect river of life rippling through spacetime. They are the kings of their forest, not the court jesters.

Another magic place is the Nourlangie Rock in Kakadu. A parrot lives here which flies with equal grace but with a quite different pattern. The deep swinging wing-beats of red-tailed black cockatoos make their bodies rise and fall in the sky like corks on a wave of air, as they have for hundreds of thousands of years. In the grassier areas not far away red winged parrots slide past with yet another pattern of movement. This time the flight is fast and streamlined, but just as light, the wings lifting the bird up and onward. These patterns are in something much more powerful than pixels.

All these, together with my sense of wonder, would, in my view, be patterns in single cells. The parrots exist in dimensions that I will never experience directly. All I can see are the patterns that my brain selects from the maps it creates of the outside world. For all he knew about light, Newton had no idea how these sensations might arise. I would like to think we could work that out. When I saw these parrots I had not begun to think about the inside of my head, but the patterns are built into my memory. I can see them now. Movement does not take time, it is part of a single thought. Maybe there is more to grasp about awareness than just richness, maybe we can begin to crack the code.

If each cell is separately conscious of patterns, which cells are picking out which patterns and passing them on to other cells? Which cells recognise beauty? Which cells decide to lock patterns away in memory? People have been asking questions like for a century or more, but not on the basis that each cell has its own view on the world. Awareness in each cell does not necessarily change the way these questions should be answered. It does change the way the questions are posed and it may be a change that allows us to arrive at a sensible answer, rather than chasing round and round in circles.

Paintings

Although it would have been difficult for me to be luckier in the time and place of my birth, one thing that saddens me is that my lifetime has coincided with the almost complete collapse of the art of painting. Painting and music are the doing I exploring by reproduction the patterns of the receiving me. What has gone so wrong with our cultural life that painting has come to an end? How, at the millennium, can Robert Hughes survey the visual arts since the time of his classic series, the Shock of The New, and see almost nothing to comment on.

Did painting die because of photography? No, most of the pictures most people really like were painted after photography was invented. Copying reality was never the point. We treasure Dürer, Boticelli, Michaelangelo, Velasquez, Goya, or Corot because the story in the image is part of the story of a person. We are looking at a person, we are hearing what the listeners to that person's story heard hundreds of years ago. That is what sets apart the best paintings. Reliving the dialogue between me and I.

Have television and cinema killed painting? Good pictures have always been inaccessible to children; appreciation needs a trained eye. The patterns in a static image take time to learn to extract. Maybe painting is just unfamiliar to a generation reared on moving images. Maybe our language of perception has shifted for good.

Or have all the pictures been painted? Have all the symphonies been composed? Great music from the Twenty Four Preludes and Fugues to Cole Porter has often sprung from new technologies and Venetian art and Impressionism drew on new materials. Maybe the new techniques lie somewhere other than in a brush and a board. But painting has also flourished through inspiration alone. Giorgio Morandi painted jars, apparently as a million others painted jars, but with the difference that he was painting Giorgio Morandi, and other friendly jars in his space – have a look. The dialogue between me and I is transposed to the simplest of inanimate protagonists. There are new masterpieces for every new master.

A painting is a story of a personal inside story, its listeners and their tiny contributions to the flow, frozen in space and time. And the stories come in all shapes and sizes. Take a look at Monet's late landscapes, like the Haystacks series. I am fairly sure Monet is playing a gentle game with the viewer. The canvas is heaped with paint as if flung on at the moment of the sun setting. However, close inspection suggests that the picture bears no relation to these heaps of paint. It has been painted on top with no more paint than necessary. Monet wanted to imply abandon, but the real picture was painted with icy precision and he knew that the people that knew it would let him get away with it. It works so perfectly: a story within a story.

Cézanne and Seurat wanted to make things more real than real, by painting the idea, the perception, as well as the impression and its sensations. Their pictures are almost neurophysiological experiments, but no less wonderful for that. Seurat's exploration of raw sensation from light mixing with perceptions is enough to provide a lifetime of amazement. If you want a lesson in the language of visual perception; lines, colours, forms, auras, mirages, dreams, forget brain journals and go to the Kröller-Müller Collection. No two pictures are alike, each a chapter of a unique and brief personal story, at once distant and immediate. Not that we hear Seurat's inside story as his listeners heard it, but we can hear a version of it, a story prepared to risk everything for extreme experiments in seeing.

And why is it that van Gogh's sunflowers, chairs and boots seem to touch almost everyone? He seems to have found a language that we all share. Ironically, he saw himself as painting human emotion directly, but most of us are simply grabbed by the clarity of the sensations and perceptions, with the inner story coming later. Again, if we want to create a science of the language of listeners in the brain, I suspect van Gogh tells us more about the building blocks than a library full of psychological data.

All these paintings tell us new things about the inside of our heads, not just how stories describe things but how stories describe stories. Can a story describe a story? Surely, even if there are a billion cells each in my head, there must be one agent, one 'soul', which is in charge, which controls our behaviour. It would be easy for me to think that there is one controlling soul which is writing this book, deciding which words to put down and which ideas to develop. But it cannot be. At least Daniel Dennett and I agree on that.

I have noticed a rather odd thing about the physical act of writing. If I write with a pen or type I make few spelling mistakes. If I write in big letters on a blackboard with chalk or a dry wipe pen I make spelling mistakes all the time. It seems that my automatic spelling control is linked to small movements of my fingers and not to big hand movements. There are a lot more different things in charge of what I do than I might think. I can catch my storytellers out trying to pretend there is a single conscious agent, but there is no one agent in charge – basic neuroscience tells us as much.

The question of who is in charge of how we behave makes me think of two particular successors to the post-impressionists, Bonnard and Bacon. For me Bonnard is the next great artist to use light and colour to explore the inner aspects of people's stories. Bonnard's painting is particularly about the private space of his house and the way women and children, and especially his wife, make that space a space of stories. His style of painting is dubbed Intimism, yet we do not see people's emotions directly. They are not usually full face, but more often slightly head down or masked by a shadow, rather still, apparently uncommunicative. There are always clues to the presence of the viewer, Bonnard, but he is not acknowledged by his subjects. Yet, taken with his modest self-portraits, these pictures seem to tell us more about Bonnard's inside story and how he saw other people's stories than almost any other twentieth century artist.

Like van Gogh, Bonnard paintings are about what he loves, but also about what is missing, what we cannot reach. They are telling us that there are things people want to say but for which words are no use. They are telling us that some versions of our inside story may not use words at all and in many ways these may be the more important parts of us. Perhaps the 'unconscious' mind is in some ways simply the wordless mind, equally valid, equally aware, but permanently inaccessible to these pages.

I see Francis Bacon as the outstanding artist of the last century. He is difficult to like. Many people see his art as grotesque but it is unlikely that his objective was to shock with horrible things for the sake of it. He describes the ideas or perceptions of personal stories in a language of raw sensations that refuses to give way to sentimentality. In a more strident way, often full-face, he describes what he sees that he loves but also how incomplete and fragmented that is. Cecil Collins said to me that Bacon's weakness was his ambiguity but for me the ambiguity and incompleteness are his strength. The patterns that build our real life stories are patchy, broken up and unpredictable. Bacon's lifestyle indicated that he thought that the rules of society made no sense. His pictures are telling us that the rules of what we see around us do not make much sense either. His distorted images show that everything is partial, transient and only given meaning as a thing or a story in passing. I suspect he sensed we are also like that inside. There is no one agent in charge. Sanity and madness are both based on illusions. The choice is not as simple as one might think.

Even the '50s and '60s were rich in artists who used perceptions and sensations to great effect. Gorky, De Kooning and Motherwell found deeply moving shapes that must touch on qualia embedded in some Jungian library of archetypes. Blake and Hockney transported personal human stories into new settings, playing with the way people see each other. Then everything stopped, bar a few like Lucien Freud. How does one explain such a profound cultural shift?

Most attempts to explain art, so beloved of connoisseurs and historians, are charades. Trying to explain things whose value lies in being unexplainable is doomed to failure. However, the attempt might just be allowed as a way to encourage people to see a visual art that tells us about inside stories in a way that words cannot as legitimate and not to be afraid to return to it. Maybe it will re-emerge with a change in cultural fashion. Maybe a new view of who is listening to the story might help that along.

People

I have considered in what way the idea that every cell is aware separately might affect the way we look at other people. Perhaps it does not matter. When communication runs well we are not in great need of understanding of the inside of either our own heads or those of others. But when communication is less than one would like one starts to feel that it might be helpful to understand more.

My garden is decked out with virtual statues of people who have paved the way for such understanding. I could have the dramatists but I prefer to have the scientists who not only laid the foundations, but also understood the enormity of the task of building a reliable theory for the inside as well as the outside world. The statues are not standing motionless in a corner but follow the movement of the trees and the people on the grass, statues of Leibniz, Newton, Maxwell, and Einstein, people who allowed us to understand movement and interrelationships. Feynman is there, but in such a mercurial form that even a virtual statue would be a poor description. And so are Spinoza and Darwin, who taught us that what is is what is likely to be so.

Of these people, for me the coolest is Leibniz. I knew virtually nothing about him before I started thinking about the mind/brain problem. He was an expert in law, mathematics, physics, philosophy and anything else you care to mention. He and Newton both developed the calculus that underlies almost any maths other than adding up a shopping bill. However, his real strength is his analysis of what exists. Like others he saw that physics needed a layer underneath; metaphysics. What sets him aside is that he devised metaphysical rules with the precision of physics that include things which physics now seems explicitly to require. He gives little idea how he came to his conclusions, but the implication is that he had considered every other possibility and found them impossible.

It is such a pity that it is so easy to go through life without sharing something of these unusual people's stories. If they were all alive today I would not be writing this book. They would have pointed out the answer decades ago.

Science has made many attempts to satisfy our need to understand each other better. Freud brought the 'unconscious' to attention and others have elaborated on his theme. Sadly, attempts of therapists to use these ideas to help seem as often as not to backfire. Basic research into the chemical and electrical processes in the brain have identified the nuts and bolts that hold our stories together, but there is still a huge rift between the two approaches. Perhaps an understanding of the truly colonial nature of our brain cells' existence may provide a better basis for understanding the complexities and inconsistencies of our behaviour.

Communication and understanding would seem to be the basis of our success. Yet this is not as simple a relation as it might seem. It is almost as if evolution is ambivalent about these assets. Understanding of human motives could give a competitive advantage but it might also induce despair at the pointlessness of it all. A little knowledge might be a dangerous thing but it might also be a very useful compromise.

The chief barrier to understanding each other seems to be what I call mind set. This is not so much belief as an inability to change belief. There must be reasons for it. It is as if our continuing belief in our inside stories is essential for our success, but carries with it a tendency to hold to things that close us off or even drag us to destruction. Moreover, it seems possible, and not unusual, to hold to two opposite beliefs at the same time. In my clinic I constantly find people carry opposite beliefs about illness and how it will affect them. Rather than changing a belief we collect another one.

What is most disturbing is when someone one has found easy to communicate with suddenly seems to take on a mind set that makes them inaccessible, almost as if they have lost their inside story. Such changes are most understandable if they occur with episodes of depression, a stroke or with the failing of the mind with age, but often there is no apparent reason. The person one knew suddenly seems to have gone. The narrator inside seems to be telling a different, closed, story. And to some degree this seems to happen to many people. The physiologist Thomas Lewis thought hardening of the arteries stopped people thinking anything new after 45. Fortunately that has not been my experience. But things do change, probably not because of arteries but because mind set patterns come to dominate the acquisition of new ideas. What I find puzzling is when this seems to come on so quickly for no obvious reason. Perhaps mindset is self-spreading, once set up in one part of the story rapidly spreading to the whole. Perhaps without the mindset of having a single awareness we might have a better chance of retaining the flexibility of spirit that keeps us close.

Sudden changes in mind-set sometimes come with a new quest for something unrealistic, something that leads to a closing off, not an opening. Famously, Fred Hoyle became interested in viruses from outer space. Linus Pauling became obsessed with vitamin C. Francis Crick decided to unravel consciousness. And here I am writing this book. So by now you need to have worked out whether this is a book about an idea of serious merit or the result of a moderately successful biomedical scientist who has started to 'zone out'. Life does not come with the answers upside down on the back page.

One reason why I do not think that I am a middle-aged scientist who has gone off the rails is that I have always been like this. The seed of this idea came when I was 16; it just got buried for a while. My original aim was to write a book for anyone who has been 11 years old because it is at 11 that we acquire a sense of self. I suspect I have stayed 11 ever since. I have never seen myself as part of an adult establishment. Why grow up? Why lose the ability to use an alethiometer to become embalmed in a grown-up certainty that passes for fulfilment.

Beth is now 15. She knows who she is, or at least she thinks she does. She can see things I can no longer see without glasses and hear grasshoppers singing that are now for me just a whisper. Her world is as rich as it can be. She may not yet see some things as clearly as I, but that means she has not been brainwashed to fit things together the wrong way. I would like to think that she will see into gardens I shall never see. The interesting bit of our story is going through doorways. Even if we decide to come back out we have had a look.

So Beth, how do you get to look into a Secret Garden? Alfred North Whitehead (1978) has some quite good advice on this. My précis would be: First of all, never believe anything anyone says unless you are sure you see the sense in it. Secondly, follow Sherlock Holmes, rule out the impossible, but be sure that you don't rule out a lot of other things by mistake. Thirdly, work on the basis that everyone may be very right about some things in their own way, however incompatible their views may seem. Aristotle, Plato, Descartes, Newton, Hume, Berkeley, Locke, Leibniz, Spinoza, Schopenhauer, Kant, James, Russell, Turing, Dennett, Searle, Penrose, Chalmers, Velmans, Baars, Strawson, Seager, Griffin, Crick, Koch, Humphrey, Vitiello, Pribram, are all right in their way.

They just missed one thing out. We are colonies, given the story of a single being by the elision of the specious present.

Bibliography

Berman, D. (2001) Berkeley. In: *The Great Philosophers* Monk, R. and Raphael, F. Editors (London, Phoenix)

Bieberich, E. (2002) Recurrent fractal neural networks. *Biosystems* 66, 145-64.

Bohm, D. and Hiley, B. (1995) *The Undivided Universe* (London, Routledge).

Brownell, W.E. (1990) Outer hair cell electromotility and otoacoustic emissions. *Ear Hear.* 11(2), 82-92.

Burnet, F.H. (1994) *The Secret Garden* (New York, Dover).

Cajal (1952), *see:* Ramón y Cajal

Chalmers, D. (1995) Facing up to the problem of consciousness. *Journal of Consciousness Studies* 2(3), 200-219.

Chomsky, N. (2000) *New Horizons in the Study of Language and Mind* (Cambridge University Press).

Cottingham, J. (2001) Descartes. In: *The Great Philosophers* Monk, R. and Raphael, F. Editors (London, Phoenix)

Crick, F. (1994) *The Astonishing Hypothesis* (New York, Scribners).

Crick, F. and Koch, C. (2003). A framework for consciousness. *Nature Neuroscience* 6, 2, 119-126.

Dennett, D. (1991) *Consciousness Explained* (London, Penguin).

Feynman, R. P. (1985) *Q.E.D.* (Princeton University Press).

Frith, C. and Frith, U. (1999) Interacting minds – a biological basis. *Science* 286, 1692-1695.

Fröhlich H. (1968) Long range phase correlations in biological systems. *Int. Journal of Quantum Chemistry* 2, 641-649.

Gazzaniga, M. (1998) *The Mind's Past* (Berkeley, University of California Press).

Goldberg, S. (2006) *Consciousness* (Miami, Medmaster)

Glynn, I. (1999) *An Anatomy of Thought* (Oxford University Press).

Hameroff, S. (1994) Quantum coherence in microtubules. *Journal of Consciousness Studies* 1(1), 91-118.

Hardcastle, V.G. (1995) *Locating Consciousness* (Amsterdam, Benjamins)

Hebb, D.O. (1949) *The Organisation of Behaviour* (New York, Wiley).

Heimburg, T. and Jackson, AD. (2005) On soliton propagation in biomembranes and nerves. *Proc Nat Acad Sci U.S.A.* 102, 9790-5.

Hodgkin, A.L., and Huxley, A.F. (1952) A quantitative description of membrane current and its application to conduction and excitation in nerve. *Journal of Physiology* 117, 500-44.

Hume, D. (1999) *An Enquiry Concerning Human Understanding* (Oxford University Press).

Humphrey, N. (1992) *A History of the Mind* (New York, Springer).

Iwasa, K., Tasaki, I. and Gibbons, R.C. (1980) Swelling of nerve fibres associated with action potentials. *Science* 210, 338-9.

James, W. (1890, reprinted 1983) *The Principles of Psychology* Chapter VI. (Cambridge MA, Harvard University Press).

Kandel, E.R. (2001) The molecular biology of memory storage: a dialog between genes and synapses. *Bioscience Reports* 21(5), 565-611.

Koch, C and Segev, I. (2000) The role of single neurons in information processing. *Nature Neuroscience* supplement 3, 1171-1177.

Kumánovics, A., Levin, G. and Blount, P. (2002) Family ties of gated pores; evolution of the sensor module. *FASEB Journal* 16, 1623-29.

Le Fanu, J. (1999) *The Rise and Fall of Modern Medicine* (London, Little Brown).

Levy, N. (2005) Libet's impossible demand. *Journal of Consciousness Studies* 12, 12, 67-76.

Liberman, E.A. and Minina, S.V. (1996) Cell molecular computers and biological information. *Biosystems*. 38, 173-7.

Libet, B. (2002) The timing of mental events. *Consciousness and Cognition*, 11, 291-9.

Lodge, D. (2001) *Thinks* (London, Penguin).

London, M. and Hausser, M. (2005) Dendritic computation. *Annual Review of Neuroscience* 28, 503-32.

McFadden, J. (2002) Synchronous firing and its influence on the brain's electromagnetic field. *Journal of Consciousness Studies* 9 (4), 23-50.

Mountcastle (1968) *Medical Physiology* (St Louis, Mosby)

Nimchinsky, E.A., Gilissen, E., Allman, J.M., Perl, D.P., Erwin, J.M. and Hof, P.R. (1999) A neuronal morphologic type unique to humans and great apes. *Proc Natl Acad Sci USA* 96(9), 5268-73.

Noë, A. (2002) Is the visual world a grand illusion? *Journal of Consciousness Studies* 9 (5-6), 1-13.

Panksepp, J. (2002) The primordial self. In: *Consciousness* Carter, R. (Ed.) pp 186-188. (London, Weidenfield and Nicholson).

Penrose, R. (1994) *Shadows of the Mind*. (Oxford University Press).

Perez-Otano, I. and Ehlers, M.D. (2005) Homeostatic plasticity and NMDA receptor trafficking. *Trends in Neuroscience* 28(5), 229-38.

Petrov, A. (1999) *Lyotropic State of Matter* (Amsterdam, Gordon and Breach).

Pockett, S. (2002) Difficulties with the electromagnetic field theory of consciousness. *Journal of Consciousness Studies* 9 (4), 51-56.

Polsky, A., Mel, B.W. and Schiller, J. (2004) Computational subunits in thin dendrites of pyramidal cells. *Nature Neuroscience* 7, 621-7.

Popper, K. (1984) *Conjectures and Refutations* (London, Routledge and Kegan Paul)

Preston, J. and Bishop, M. (2002) *Views Into The Chinese Room* (Oxford University Press)

Pribram, K.H. (1991) *Brain and Perception* (New Jersey, Lawrence Erlbaum).

Pullman, P. (1995) *Northern Lights* (New York, Alfred A Kopf).

Quiroga, R.Q., Reddy, L., Kreiman, G., Koch, C., Fried, I. (2005) Invariant visual representation by single neurons in the human brain. *Nature* 435(7045), 1102-7.

Ramón y Cajal, S. (1952) Structure and connections of neurons. *Bulletin Los Angeles Neurological Society* 17, 5-46.

Russell, B. (2002) *Our Knowledge of the Outside World* (London, Routledge).

Sacks, O. (1986) *The Man Who Mistook his Wife for a Hat* (London, Pan).

Sacks, O. (2000) *Seeing Voices* (London, Vintage)

Scott, A. (2003) *Non-Linear Science* (Oxford University Press).

Scruton, R. (2001) Spinoza. In: *The Great Philosophers* Monk, R. and Raphael, F. Editors (London, Phoenix)

Seager, W. (1995) Consciousness, Information and Panpsychism. *Journal of Consciousness Studies* 2 (3), 272-288.

Searle, J. (1997) *The Mystery of Consciousness* (London, Granta Books).

Sevush, S. (2006) Single-neuron theory of consciousness. *Journal of Theoretical Biology*, 238, 704-25.

Spence, S. (2006) The cycle of action: commentary on Young. *Journal of Consciousness Studies* 13, 3, 69-72.

Strawson, G. (1999) The 'self' and the SESMET. *Journal of Consciousness Studies* 6 (4), 99.

Vitiello, G. (2001) *My Double Unveiled* (Amsterdam, John Benjamins).

Whitehead, A.N. (1978) *Process and Reality* (New York, The Free Press)

Williams, S. (2004) Spatial compartmentalization and impact of conductance in pyramidal neurons. *Nature Neuroscience* 7, 961-7.

Woolhouse, R.S. and Franks, R. (1998) *G.W. Leibniz, Philosophical Texts* (Oxford University Press).

Zhang, P-C., Keleshian, A.M. and Sachs, F. (2001) Voltage-induced membrane movement. *Nature* 413, 428-31.

Index